www.brookscole.com

www.brookscole.com is the World Wide Web site for Brooks/Cole and is your direct source to dozens of online resources.

At *www.brookscole.com* you can find out about supplements, demonstration software, and student resources. You can also send email to many of our authors and preview new publications and exciting new technologies.

www.brookscole.com
Changing the way the world learns®

SECOND EDITION

Experiments in Biochemistry: A Hands-On Approach

A Manual for the Undergraduate Laboratory

Shawn O. Farrell
Colorado State University

Lynn E. Taylor
Colorado State University

THOMSON

BROOKS/COLE

Australia • Canada • Mexico • Singapore • Spain • United Kingdom • United States

THOMSON

BROOKS/COLE

Experiments in Biochemistry: A Hands-On Approach, second edition
Shawn O. Farrell, Lynn E. Taylor

Publisher: *David Harris*
Technology Project Manager: *Donna Kelley*
Marketing Manager: *Amee Mosley*
Project Manager, Editorial Production: *Belinda Krohmer*
Art Director: *Rob Hugel*
Print Buyer: *Lisa Claudeanos*
Permissions Editor: *Kiely Sisk*

Production Service: *Matrix Productions*
Text Designer: *Carolyn Deacy*
Copy Editor: *Betty Duncan*
Cover Image: *©Tom Grill/Corbis*
Cover Printer: *West Group*
Compositor: *Cadmus*
Printer: *West Group*

For more information about our products, contact us at:
Thomson Learning Academic Resource Center
1-800-423-0563

For permission to use material from this text or product, submit a request online at
http://www.thomsonrights.com.
Any additional questions about permissions can be submitted by email to
thomsonrights@thomson.com.

Library of Congress Control Number: 2004113417
ISBN-13: 978-0-495-01317-4
ISBN-10: 0-495-01317-X

Thomson Higher Education
10 Davis Drive
Belmont, CA 94002-3098
USA

Asia (including India)
Thomson Learning
5 Shenton Way
#01-01 UIC Building
Singapore 068808

Australia/New Zealand
Thomson Learning Australia
102 Dodds Street
Southbank, Victoria 3006
Australia

Canada
Thomson Nelson
1120 Birchmount Road
Toronto, Ontario M1K 5G4
Canada

UK/Europe/Middle East/Africa
Thomson Learning
High Holborn House
50/51 Bedford Row
London WC1R 4LR
United Kingdom

Latin America
Thomson Learning
Seneca, 53
Colonia Polanco
11560 Mexico
D.F. Mexico

Spain (including Portugal)
Thomson Paraninfo
Calle Magallanes, 25
28015 Madrid, Spain

Introduction to the Text

For Teachers and Students *Experiments in Biochemistry* was written to complement a wide variety of biochemistry lecture formats. At Colorado State University, we have two levels of biochemistry lecture, with the choice based on career goals and major requirements. I had used Mary Campbell's *Biochemistry* for both lecture classes and found that it contained the best level of detail and simplicity for those students. For that reason, this manual was written to complement such a text as closely as possible.

Some lab classes are organized so that each experiment is independent. Often these classes meet only once a week and must provide the greatest overview possible in 9 to 15 weeks. Other classes, perhaps for biochemistry majors, meet for longer times and more often. Such classes may wish to have experiments that flow more smoothly from one to another. For this reason, the experiments in this manual are divided into two types. For many sections, an experiment can stand alone, followed by a similar experiment that can be part of an overall plan. These comprehensive experiments usually have an experiment number followed by the letter *a*. For most of the *a*-type experiments, it is assumed that the student will have retained the sample collected during the previous experiment.

The model system for the comprehensive experiments is lactate dehydrogenase (LDH). An entire semester can be organized around the isolation, purification, and characterization of beef heart LDH, followed by the cloning and expression of recombinant barracuda LDH-A in bacteria.

The methodology given is often specific for the types of equipment that we use in our lab, but any of these experiments could be optimized for similar but different devices.

Most of these experiments have been designed and improved by many of my students and teaching assistants, and we are very happy with the results. All experiments are proven to work, and they can be completed in a normal lab period by a competent and prepared student.

We are most excited by the new Experiment 12, which involves cloning the barracuda LDH gene into an expression vector and making a fusion protein that can be purified by affinity chromatography. My upper-division lab class worked out the details on this experiment during the spring 2004 semester, and the results were tremendous.

We hope that you enjoy doing these experiments as much as we have enjoyed designing them.

Suggested Order for Comprehensive Purification of LDH

1. Purifying LDH (Comprehensive Version), Experiment 4a

2. Purifying LDH with Ion-Exchange Chromatography, Experiment 5a

3. Affinity Chromatography, Experiment 6a

4. Gel Filtration Chromatography, Experiment 7a

5. Protein Concentration of LDH Fractions, Experiment 3a

6. Enzyme Kinetics of LDH, Experiment 8a

7. Native Gel Separation of LDH Isozymes (Comprehensive Version), Experiment 9a

8. SDS-PAGE (Short Version), Experiment 9c

9. Western Blot of LDH, Experiment 10a

10. Cloning and Expression of LDH (or parts thereof), Experiment 12

11. Polymerase Chain Reaction of LDH, Experiment 13

Acknowledgments

I would like to thank Mary K. Campbell, coauthor of *Biochemistry*, and Brooks/Cole Publishers for permission to use figures from the aforementioned text.

Special thanks go to Bio Rad Laboratories, Fisher Scientific, Milton-Roy, Calbiochem, Rainin Instruments, and Sigma Chemicals for permission to use figures from their instruction manuals.

The molecular biology portions of this manual would not have been possible without the generous assistance of Dr. Linda Holland from the University of California at San Diego.

I would also like to thank the following reviewers for their suggestions and improvements for this edition:

Dr. Pamela K. Kerrigan, College of Mount Saint Vincent/Manhattan College
Edwin A. Lewis, Northern Arizona University
Ruel McKnight, SUNY Geneseo
Dr. R. Marshall Werner, Lake Superior State University
Jacqueline Whitling, Lock Haven University

Technical Help

An *Instructor's Manual*, which includes instructor's guidelines to equipment and reagents, is available to adopters of this book through your local Brooks/Cole representative. Technical help concerning material in this manual is available from the author. Feel free to contact Dr. Shawn Farrell at the following email address and phone: sfarrell@usacycling.org 719-229-0732

Contents

Introduction to the Biochemistry Lab

Welcome to the biochemistry laboratory! Biochemistry is a fascinating subject that overlaps many other scientific fields. The techniques that you learn in a biochemistry lab will be applicable to all life sciences across a broad range of professional interests. Many courses in biology, microbiology, chemistry, botany, zoology, food science, and nutrition have laboratories that include large sections of biochemistry experiments and techniques. Whether your professional interest is in working in a lab as a researcher, going to graduate school, or entering a professional degree program such as medicine or veterinary medicine, you will need the skills and knowledge of biochemistry.

Objectives of the Biochemistry Laboratory

A biochemistry laboratory has several objectives. These include learning

- Physical skills and techniques of modern experimental biochemistry.

- How to think scientifically, think independently, and do the requisite calculations.

- The theory behind the techniques and the biochemical pathways.

Physical skills involve using a pipet correctly, using a pH meter properly, balancing centrifuge tubes, loading chromatography columns, setting up electrophoresis equipment, and so on. The techniques refer to the actual types of experiments; each experiment has a specific goal and a proper time and place for its use. The skills and techniques are the most important part of this experience. Why? Because when you graduate, if you can list on your résumé that you have done, for example, HPLC, SDS-PAGE, and agarose gel electrophoresis of DNA, you will immediately impress potential employers. If you go into graduate school with a better-than-average understanding of these techniques, then you will avoid needless repetition in your first-year graduate courses. This puts you a step ahead and starts your real research faster. If you immediately go into a field such as medical technology, then these techniques will be your livelihood.

Thinking scientifically and doing calculations is also an important part of laboratory science. You will never be just a pair of hands doing work; planning future experiments and analyzing the data from past ones will always need to be done. If you understand why things work the way they do, you will plan your experiments correctly. This saves time and money in the long run. It would be a tragic shame if you did a brilliant, elegant experiment and then got the wrong conclusions because of a simple math error. There are always calculations at the end of an experiment, and they are very, very important. Never say or think, "That didn't matter. It was just a math error. I understood the important stuff." In the medical field, which many of you may aspire to, simple math errors can be fatal. In the graduate laboratory, simple math errors can cost several month's work and your boss several months of supplies.

The basic knowledge of biochemistry and the related pathways are learned effectively in a lecture course, but hands-on experience in the lab reinforces that knowledge. That is why all experiments in this manual revolve around some real biochemistry. Because you may not be taking this lab concurrently with a lecture course, I have tried to give all the necessary background for each experiment, but it is assumed that you have a basic biochemistry text, such as *Biochemistry* by Mary Campbell and Shawn Farrell, or *Biochemistry* by Reginald Garrett and Charles Grisham. The theory behind the techniques is important so that you know how to apply the techniques in a new situation and plan experiments from the beginning.

Chapter Format of Experiments in Biochemistry

Each chapter is organized roughly according to the following areas:

Topics. This is what you should get out of reading each chapter and doing each experiment. If you have done the experiment, read the material carefully (perhaps more than once), and still do not have a thorough understanding of the concepts, then one of us has not done our job well enough.

Introduction. This gives the background material that I would normally present in a lecture course about the same topic. Sometimes the background may be more intensive than you need for the fairly simple technique that follows it, but it is better to be overthorough in this case. Between this manual and a basic biochemistry text, you should be able to complete all experiments and answer all questions without spending time doing extensive research from journals. The most important concepts are also summarized in the box **Essential Information.** Problem-solving skills are demonstrated with the **Practice Sessions.** Some chapters have the section **Expanding the Topic,** which gives information that is not necessarily essential for completing the experiment but goes into greater depth on some of the fine points.

Prelab Questions. These are specific questions pertinent to the experiment that you are about to do. I include them for a couple of reasons. First, if you have read the material, you should know how to answer these questions. If you do not, then you need to go back over it again. Second, if you do not know the answers to the questions, you are about to waste valuable lab time reading when you should be doing the experiment. That will put you behind, and you and your lab part-ner will probably be getting out of the lab late. Third, your instructor may ask that you turn these questions in prior to the lab as part of your lab grade. I always do.

Experimental Procedures. This section gives you the materials and meth-ods you will be using for that experiment in a step-by-step procedure. *Always, and I repeat, always read all procedures before beginning anything!*

Analysis of Results and Questions. This is where you report your results and do the calculations for your data.

Additional Problem Set. Here you will find some more problems that your instructor may ask you to do, or you may just use them for prac-tice. You should be able to solve all problems with just this manual and occasionally a basic biochemistry text.

Webconnections. The Brooks/Cole web site maintains links for many helpful sites. You can also access information specific to this manual by going to www.brookscole.com and selecting the link for *Experiments in Biochemistry.*

References and Further Reading. These tell where much of the infor-mation in this manual came from. They are listed for professional eti-quette reasons, not as required reading. Also listed are books and articles about particular topics in case you want to know more.

Chapter 1

Biochemistry Boot Camp

SURVIVAL IN THE BIOCHEMISTRY LAB

TOPICS

Introduction

In this chapter, we discuss some fundamentals of laboratory science. This material is often overlooked by instructors who, correctly or incorrectly, assume that you have learned it well in other courses. It could be argued that the material presented here is the most important, however, due to the shear magnitude of the repetition. You will constantly be doing calculations involving units, concentrations, and dilutions. You will be presenting your data using tables and graphs. You will do hundreds of pipettings during a semester. If this material is not mastered, you will pay a heavy price all semester. Read this chapter thoroughly, even if you think you already know it. You may just find an informational jewel you had not expected.

1.1 Lab Safety

A lab can be a dangerous place if you are not careful. This potential danger comes from several sources. First, the very nature of the chemicals and equipment used may be hazardous. We try to minimize use of hazardous materials, but some modern techniques in biochemistry require that we use them. Hazardous chemicals are noted in the "Materials" section of the experiments. Some equipment may be dangerous due to moving parts, heat, or potential electric shock. Only by using the equipment correctly, as instructed, will you be sure that you are safe.

Second, glassware used in a lab is always considered dangerous because it breaks when dropped or mishandled. Flying glassware may come from anywhere, so you may not be the one who makes the mistake, but you may be the one who pays for it. Proper clothing and eye protection is the only sensible way to protect yourself against most common lab accidents.

Third, most accidents are caused by carelessness. A student who is mentally prepared to undertake the lab, has studied the material, and understands the procedures is much less likely to make a mistake that could injure someone. Simple mistakes like careless washing of glassware can be dangerous if water ends up on the floor and isn't cleaned up. Note and observe the following lab procedures.

Things You Should Do

1. Always use some form of protective eyewear. The most reliable are certified lab goggles that protect from the sides as well as the front. Contact lenses can be a problem in a lab because tears do not wash out things spilled in the eye when a contact lens is present. Eyewash stations are not efficient if you are wearing contacts. The American Chemical Society, however, has recently removed its recommendation against contact lenses in the lab. Evidence shows that contacts are not dangerous **IF** proper protection (goggles) is used.

2. Be aware of what chemicals you are using. Wear gloves when using toxic chemicals. Remember that, although you are not using a dangerous chemical, the student beside you may have spilled one right next to you.

3. Wear proper clothing in the lab. The lab is a good place for long sleeves, long pants, closed-toed shoes, and standard-lens glasses. It is a bad place for short sleeves, shorts, and sandals. Don't even think about going into a lab barefoot.

4. Familiarize yourself with the layout of the lab. Do you know where the fire extinguishers are? Where are the eyewash stations? Where is the first aid kit?

5. Label all reagents that you bring to your bench with tape.

Things You Should Never Do

1. Never eat, drink, or smoke in the lab. Although you may see more advanced scientists doing the first two in their labs, it is a bad idea in a teaching lab. There may be 200 students using that lab in a week, and the instructor cannot control what all of them are doing. If you need to drink often, then bring a water bottle but leave it outside. Nobody will mind it if you step outside to eat or drink, but bringing any food or beverage into the lab, even if sealed, is a potential danger.

2. Never use mouth suction on glass pipets to draw up a solution. Even if you think the solution is a harmless buffer, the pipet may be contaminated with something hazardous. Use pipet pumps and bulbs to draw up solutions.

3. Never work alone. Someone else should always be in lab with you when you are working.

Additional Lab Courtesy

Although lab etiquette is not strictly a safety issue, you should follow it as well. The following items will make your lab run more smoothly and lead to efficient transitions between lab sections.

1. Never stick your personal pipets into a community reagent bottle. If your pipet is dirty, you will contaminate the supply for the whole class. If you need 10 mL of a reagent and a 1-L bottle is in a community reagent area, use a small beaker and pour in about 10 mL. That way, if your beaker is dirty, you will have contaminated only your own supply.

2. Never take a community reagent back to your own bench. One of the most frustrating things that can happen in a lab is not finding something you need.

3. Don't use more of the chemical than you need. Students often take chemicals from the community area before they have the slightest clue about how much they need. They look at the size of the community reagent bottle and try to guess from there. Just because the reagent is in a 1-L bottle doesn't mean that you should take 100 mL of it. You might only need 2 mL of it. The 1-L bottle may be the total supply for all sections of the lab for the week.

4. Always clean up your lab area and any equipment and glassware that you used. The next class may need to use the same materials. The job is not over until the lab is clean and the equipment is ready for the next class.

1.2 Scientific Notation

You will probably see many numbers written in many different ways during the course of your lab. The idea is to accurately and clearly communicate numerical information, and nothing is necessarily wrong with any system of numbers that does that. However, some traditional ways of handling numbers are used in science. One of these is called scientific notation.

Any number in **strict** scientific notation starts with one nonzero digit followed by a decimal point and some other numbers. That is followed by an exponent that tells to what power to raise it. Thus, the number 623 would be written 6.23×10^2. The number 0.0456 would be written as 4.56×10^{-2}.

When not using strict scientific notation, avoid starting a number with just a decimal. Thus, rather than writing .453, write 0.453. When you label a test tube for storage, sometimes the ink fades. It always seems like the decimal point fades faster than the numbers. The beauty of using strict scientific notation is that a fading decimal point wouldn't matter because you would always know that the decimal point belonged after the first number.

1.3 Significant Figures

When you take a measurement or do a calculation, you can only rely on a certain amount of the information that you get from it. Analysis of significant figures is our way of determining how reliable the numbers are. Nowadays, many students use calculators or computers for everything, often with humorous results.

If you have a standard ruler with inches on one side and centimeters on the other and ask me to measure the height of your 250-mL beaker, what will you think if I tell you it is 8.423587763 cm? You might say, "Thanks, Mr. Spock, but 8.4 cm is sufficient," but you will certainly see that I cannot measure something to that many decimal places when the smallest division on the ruler is 1 mm, or 0.1 cm. However, you might easily make a very similar mistake when you do a calculation. If you measure a cube to be 9.6 cm on a side and want to calculate the volume in cubic centimeters, you might multiply $9.6 \times 9.6 \times 9.6$ and report the volume as 884.736 cc. Students do that all the time without recognizing that they are as wrong as in the Mr. Spock example. If each side is measured to an accuracy of one decimal place, the answer cannot be given in three decimal places.

Definition of Significant Figures

So, what is a significant figure? That has to do with the accuracy and precision of the instrument being used to make the measurement. If you have a balance, you weigh out a sample of histidine, and the balance reads 1.473 g, then you have four significant figures. The last figure, 3, is probably at the limit of the machine and is an estimate. If you weigh the same sample again, it might read 1.472 or 1.474. This is like estimating between the closest marks on a ruler. Maybe a cruder balance reads 1.5 g, and you then have only two significant figures.

The number and position of zeros is often confusing to students with regards to significant figures. This is another reason to use scientific notation because there is no ambiguity when using it. For example, how many significant figures are in 0.0456? The answer is three. Zeros before the first nonzero digit have nothing to do with the accuracy; rather, they mark the place of the decimal point in scientific nomenclature. In scientific format, this number is 4.56×10^{-2}. How many significant figures are in 0.045600? In this case, there are five. The zeros following the 6 are significant and tell you about the accuracy of the measurement. In scientific notation, this number is 4.5600×10^{-2}. In scientific notation, all zeros are significant. The real ambiguity comes when you write a number like 3200. How many significant figures are in 3200? You can't really tell. You could mean 3.2×10^2, 3.20×10^2, or 3.200×10^2, which would be two, three, or four significant figures, respectively.

Significant Figures in Calculations

There are a couple of simple rules for using significant figures in calculations.

Multiplication and Division When you multiply numbers, the answer will have the same number of significant figures as the number with the fewest. For example, if you want to calculate the volume of a cube by multiplying the length of the sides, it might look like this:

$$(3.4 \, cm) \times (56.8 \, cm) \times (2.435 \, cm) = 470.2472 \qquad \text{(on our calculator)}$$

How many significant figures can you claim in your answer? The answer is two because the first number you multiplied has only two significant figures. So the answer is $4.7 \times 10^2 \, cm^3$. Division is done exactly the same way.

Addition and Subtraction Addition and subtraction are a little different. When you add strings of numbers, look at the number of decimal places to determine the accuracy of the measurement. The final answer cannot be more accurate than the least accurate number added. For example, if you are adding volumes to get a total, it might look like this:

$$(22.4 \, mL) + (3.5 \times 10^2 \, mL) + (0.543 \, mL) = 372.943 \, mL$$

How many significant figures can you claim? In this case, we can't really look at the significant figures in each term; rather, we look at decimal places. The figure with the fewest decimal places is 350. Therefore, our answer cannot go into tenths and hundredths. The true answer is 373 mL. Notice that the answer has three significant figures whereas one of the numbers added has only two.

1.4 Statistics and Scientific Measurements

As a scientist, you will deal with a great many numbers. We use statistics to help us get more meaning out of the numbers. An example any student can relate to would be test scores. If your friend tells you, "Hey, I got a 40 on the last exam," you would not immediately know whether you should be happy for him or not. How happy you would be might depend on several issues, including the total possible points for the exam, the average score on the exam, the standard deviation, and your score on the exam. You might also need to know whether you could really believe that number. Was the score added correctly? Was it a subjective or an objective score? To truly understand how that exam score relates to your friend's ability in that subject, you also have to know whether it is likely that he would score the same again on a similar exam or is that score overly high or low for some reason. When making measurements during experiments, we need to draw meaning from the numbers we see, and we often use statistics for this purpose.

If 200 students take the first exam in a class, we may list their scores as $x_1 = 85$, $x_2 = 64$, $x_3 = 98, \ldots, x_{200} = 12$. An individual number is usually designated x_i. What can we tell from these data? Students will want to know what their score means. If the professor is using a straight-scale grading system where 90–100% is an A, 80–89% is a B, and so on, then the first student (x_1) knows that she got a B. The third student knows that he got an A. The last student (x_{200}) knows that she should consider dropping the class. If the professor is using a grading curve, then students need to know some statistics to figure out how they are doing in the course.

There are two basic types of statistical measures. The first is a **measure of central tendency.** The one most often seen is the **average,** or **arithmetic mean** (x_{avg}). The average is calculated by adding all the numbers and dividing by the number of scores.

$$\textbf{arithmetic mean } (x_{avg}) = \frac{x_1 + x_2 + x_3 + \cdots + x_n}{n} = \frac{\Sigma x_i}{n}$$

Therefore, we would add the scores for all the students and divide by 200 to get the average on the first exam. If the average turns out to be 85, then the student with a 98 will feel pretty good. The student with an 85 would know that she is at the middle ground. On the other hand, if the average is only 40, then even the student scoring in the 60s will know that he has done better than most. However, to calculate their current grades on a curve, students need more information.

A second type of statistical measure is a **measure of dispersion.** It usually measures the variations in the values compared to the mean. Often such a measure is necessary to truly understand the data presented. If I am the professor of the class taking the exam and I see that, of my 200 students, 180 of them scored an 85 on the exam, with a few being higher and a few lower so that the average was still an 85, I will know that I have a very homogeneous class. Everybody did about the same with just a few exceptions. If, on the other hand, the scores are scattered all over with many students scoring below 40 and many above 90, I will know that the class is heterogeneous. I would adapt my teaching strategy differently to the two classes. When grades are assigned on a curve, it is always the mean and some measure of dispersion that determines the grade.

There are many different measurements of dispersion, and the exact reason for using one or another is beyond the scope of this text. Some of the more common ones are presented next.

$$\textbf{mean deviation} = \frac{\Sigma |x_i - x_{avg}|}{n}$$

where n is the number of samples measured and the | | means the absolute value of the difference between the individual value and the mean.

$$\text{percent deviation} = \frac{\text{mean deviation}}{\text{mean}} \times 100$$

$$\text{variance} = \sigma^2 = \frac{\Sigma(x_i - x_{\text{avg}})^2}{n - 1}$$

$$\text{standard deviation} = \sigma = \sqrt{\sigma^2}$$

The mean plus the standard deviation are the two most common statistical parameters. Students taking the exam will want to know their score in relation to the mean and the σ (sigma) because if the class is graded on a strict bell curve, it usually takes a score of the mean $+1\sigma$ to give a grade of B and mean $+2\sigma$ to give a grade of A. Most students' calculators will determine mean and standard deviation for sets of data.

Errors in Experiments

You will often read about errors and error analysis in the data reported for an experiment. There are many sources of error in an experiment, and the term is often used as a catchall to explain numbers that are not perfect. It is important to realize that statistical error may not mean the same thing as the errors that you are more familiar with. When you think of an error, you think of something like multiplying 6 times 9 and getting 42. When scientists think of error, they more often think of differences in members of a population. For example, if you measure the activity of the enzyme lactate dehydrogenase from the serum of ten different people and express it as enzyme units per milliliter of blood serum, you will not get the same number twice. You might have a range of values from 0.1 unit/mL to 26 units/mL. You could calculate the average and standard deviation and report your findings as

$$18.3 \pm 5.3$$

That is error analysis, but the error does not imply that you made a mistake in your work. This is biological error and simply reflects the individual variation in a population. You could have reported the values as mean $\pm$ variance or mean $\pm$ mean deviation. Each way would have given you different numbers.

If you attempt to pipet 1 mL with a Pipetman and you actually pipet 0.7 mL, there is error in the process. Is it caused by your inability to pipet correctly? Is the Pipetman miscalibrated? In this case, there is a *true* value that is known, so it is easier to determine the source of error. When sampling enzymes in the sera of biological organisms, it is more difficult to know the true value.

Accuracy versus Precision

A measurement is **accurate** if it gives the true value. If you attempt to pipet 1 mL of water, which should weigh 1 g, and the balance reads 1 g after you dispense the solution, your pipetting is accurate. The arithmetic mean is the usual grounds for accuracy.

Measurements are **precise** if the same measurement can be made again and again. For example, if you try to dispense the 1 mL with the pipet ten times and you dispense 0.7 mL ten times in a row, your pipetting was very precise but inaccurate. This usually draws attention for the problem away from you and onto the equipment being used. Measurements of dispersion are the usual criteria for precision.

Another parameter that can be measured in the case of pipetting is the **percent (%) error.** If you are trying to measure a known quantity, such as 1 mL of water, and you expect it to weigh 1 g, the % error will give you a relative estimate of the error of your pipet or your pipetting technique.

$$\% \text{ error} = \frac{\left| x_{\text{avg}} - x_{\text{true}} \right|}{x_{\text{true}}} \times 100$$

So, in our pipetting example, if your average had been 0.7 g when it was supposed to be 1 g, the % error would be calculated as follows:

$$\% \text{ error} = \frac{\left| 0.7 \text{ g} - 1.0 \text{ g} \right|}{1.0 \text{ g}} \times 100 = 30\%$$

1.5 Units

The international system of measurements is known as the SI (from *Système International d'Unités*) and is based on the MKS (meter–kilogram–second) system. The base units for SI are given in Table 1.1.

Many other units are derived from SI units, such as the joule ($m^2 kg/s^2$), the unit of energy. Some non-SI units are also frequently used, such as degrees in Celsius (°C), pressure in atmospheres, etc. The common volume liter is not a SI unit because volume is measured in cubic meters. A liter is actually a cubic decimeter.

TABLE 1.1 *Basic SI Units*

Length	meter	m
Mass	kilogram	kg
Time	second	s
Temperature	Kelvin	K
Electric current	ampere	A
Amount	mole	mol
Radioactivity	Becquerel	Bq

TABLE 1.2 *Prefixes for Multiple Units*

Quantity	Prefix	Abbreviation
10^{12}	tera	T
10^{9}	giga	G
10^{6}	mega	M
10^{3}	kilo	k
10^{-1}	deci	d
10^{-2}	centi	c
10^{-3}	milli	m
10^{-6}	micro	μ
10^{-9}	nano	n
10^{-12}	pico	p
10^{-15}	femto	f
10^{-18}	atto	a

Multiples of these units are most frequently used in biochemistry, and it is important that you know these. Table 1.2 gives the most common multiples. With these multiples, we arrive at units such as millimeters (mm), micromoles (μmol), etc.

Very, very few numbers in biochemistry exist without units. If you calculate the volume of something to be $12.34 \, cm^3$, your answer will be wrong if you just report the number. It will be just as wrong as if you got the number wrong. It may seem like a small point to forget a unit, but think about the situation in which a physician tells a nurse to prepare a hypodermic syringe with 10 of phenobarbital. What will the nurse prepare? Ten what? Ten milliliters? Ten milligrams? If it is 10 mL, then 10 mL of what concentration? Is the drug absolutely pure or diluted? Without units, the nurse wouldn't know what to prepare.

1.6 Concentration of Solutions

Every experiment you do will use at least one solution of a chemical. When chemicals are used as liquids, they can be either pure liquids or solutions. An example of a pure liquid is absolute ethanol. The chemical is a liquid, but every molecule in the bottle is ethanol with nothing mixed in. A **solution** contains a chemical dissolved in another liquid. When this is the case, you must not only know what the chemical of interest is but also what it is dissolved in and how much of it is dissolved. The chemical in the smaller quantity is called the **solute;** the liquid it is dissolved in is called the **solvent.** When calculating the amount of the solute dissolved in the solvent, you have determined the concentration. The concentration is very important to the biological or chemical function of a liquid. A good example is making a glass of Gatorade from powder. If you follow the directions, you would put one scoop into an 8-oz container of water. If you instead mix three scoops with 8 oz of water, you will still have Gatorade, but its concentration will be totally wrong. It will taste terrible and will not empty from your stomach properly. You might actually become dehydrated during your workout by using it.

Definition of Concentration

A **concentration** is always the ratio of an amount of a chemical divided by the total volume. Distinguishing values that are concentrations from those that are not, which we will call **amounts,** is important. Remember that amounts are additive, but concentrations are not. For example, if we put 1 g of salt in a beaker, that is an amount. If we bring the volume up to 1 L with water, then we have a 1-g/L solution, which is a concentration. If we have 1 millimole (mmol) of salt in a beaker, that is an amount. If we bring the volume up to 1 liter with water, then we have a 1 mM (millimolar) solution, which is a concentration.

If we have 1 g of salt in one beaker and 1 g of salt in another beaker and we add them together, we will know that we have 2 g of salt. **Amounts are always additive.** If we have a solution that is 1 g/L and we add it to a solution that is 2 g/L, we do *not* get a solution that is 3 g/L. In fact, the solution would have a concentration between 1 and 2 g/L based on the volumes that we added. **Concentrations are not additive.**

Percent Solutions

Solutions based on percent are the easiest to calculate because they do not depend on a knowledge of the molecular weight. Your instructor can give you a tube with an unknown white powder and tell you to make up a 1% w/v solution in water, and you can do it without knowing anything about the white powder.

> **% w/v** means percent weight to volume and has units of grams/100 mL. Therefore, a 1% w/v solution has 1 g of solute in a total of 100 mL of solution.

> **% v/v** means percent volume to volume and has units of milliliters/100 mL. Therefore, a 1% v/v solution of ethanol has 1 mL of pure ethanol in 100 mL of total solution.

You can assume that, unless told otherwise, the solvent is water for any solution.

TIP 1.3 Derivatives of molar solutions, such as mM, μM, and nM, still refer to an amount divided by liters of solution. Many students mistakenly think that a 1-mM solution means 1 mmol in 1 mL. What it actually means is 1 mmol in 1 L!

Molar Solutions

The most common types of solutions are molar solutions. A 1-M solution means 1 mol of solute in a total volume of 1 L. A 1-mM solution has 1 mmol (10^{-3} mol) in a total of 1 L of solution. A 1-μM solution has 1 μmole (10^{-6}) of solute in a total of 1 L of solution.

Remember that moles are calculated by dividing grams by the formula weight in grams per mole. For example, if we have 13 g of compound X and its formula weight is 39 g/mol, it means that 1 mol of compound X weighs 39 g. If we have 13 grams of it, we calculate moles thusly:

$$13 \text{ g} \div 39 \text{ g/mol} = 0.33 \text{ mol}$$

If we then take the 13 g and bring it to 0.5 L with water, our molar concentration will be

$$0.33 \text{ mol} \div 0.50 \text{ L} = 0.666 \text{ mol/L} = 0.67 \text{ M}$$

1.7 Dilutions

Dilutions seem to be one of the hardest concepts for most beginning biochemistry students. Even after chanting the mantra **"Dilutions are easy, dilutions are fun, dilutions make sense,"** students still struggle. Doing problems is the only way to really understand dilutions. A book such as Irwin Segel's *Biochemical Calculations* can also be a big help.

When doing a dilution, you always start with a concentration of something and add more solvent to it, thereby lowering the concentration. *Think about that!* If you do your calculations and determine that the concentration increased, you have made a mistake. Check your work each time by making sure that your concentration did in fact decrease with your dilution.

There are two simple ways of doing all dilution problems: the $C_1 V_1$ method and the dilution factor method. Both methods are essentially the same if you understand math but also have subtle differences, which you may or may not appreciate.

$C_1 V_1$ Method

This method uses the formula

$$C_1 V_1 = C_2 V_2$$

and is good for those of you who are less comfortable with math. It is very reproducible and always works, once you know how to define the variables.

The disadvantage is that it is slower and you generate worthless intermediates if you have multiple dilutions. It is, however, a good starting point.

PRACTICE SESSION 1.1   We have a sodium phosphate buffer at a concentration of 0.2 M (mol/L). Take 10 mL of the buffer and add it to 90 mL of water. What is the final concentration after the dilution?

First, to use the formula, you need three of the four variables:

C_1 = initial concentration, which in this case is 0.2 M

V_1 = initial volume, which is 10 mL

C_2 = final concentration, which is what we want to know

V_2 = final volume. **This is a potential danger point!** The final volume is 100 mL. Remember that the volume of solvent added (that is, 90 mL) is relevant only in that it usually gives us the total volume. Sometimes 10 mL + 90 mL will not equal 100 mL, but for this example we assume that it does. For the calculation, you always use the total volume, so

$$C_1 V_1 = C_2 V_2$$

$$(0.2\,\text{M})(10\,\text{mL}) = C_2 (100\,\text{mL})$$

$$C_2 = \frac{(0.2\,\text{M})(10\,\text{mL})}{100\,\text{mL}} = 0.02\,\text{M} \qquad \bullet$$

Dilution Factor Method

The dilution factor method is faster and better for multiple dilutions. We define a dilution factor to be the final volume divided by the initial volume. This is just a convention, but we use it consistently throughout this book. Therefore, in our previous example, the dilution factor is 100 mL/10 mL = 10, or equally said, a 10-to-1 dilution.

Now that we have the dilution factor, what do we do with it? Well, there are really only a couple of things we can do. We can either multiply it by C_1 or divide it into C_1. One possibility gives an answer greater than what we started with, which we know can't be right. Therefore,

$$C_2 = \frac{C_1}{D_f}$$

$$C_2 = \frac{0.2\,\text{M}}{10} = 0.02\,\text{M}$$

Can you see that the two methods really are the same thing?

The reason that so many students get answers that indicate the concentration increased with dilution is because of nomenclature. Many books refer to a dilution as a 1 to 10, or 0.1. They get that by defining a dilution factor as initial volume over final volume. There is nothing wrong with doing that, but you will have to then multiply your initial concentration by your dilution factor. Any way you do it that is consistent should work for you, but we will always be consistent and do it the way described.

Multiple Dilutions

When you have multiple dilutions, using the dilution factor method is more efficient. If, for example, we make a dilution three times, calculate the final concentration thusly:

$$D_{f_1} = \frac{100 \text{ mL}}{10 \text{ mL}} = 10$$

$$D_{f_2} = \frac{50 \text{ mL}}{2 \text{ mL}} = 25$$

$$D_{f_3} = \frac{90 \text{ mL}}{30 \text{ mL}} = 3$$

$$D_{f_{\text{total}}} = 10 \times 25 \times 3 = 750 \text{ to } 1$$

If we start with the same 0.2-M solution, calculate the new concentration as follows:

$$C_{\text{final}} = \frac{C_1}{D_{f_{\text{total}}}} = \frac{0.2 \text{ M}}{750} = 0.00027 \text{ M}$$

$$= 0.27 \text{ mM}$$

$$= 270 \ \mu\text{M}$$

The final conversion from molar to millimolar to micromolar will eventually be simple for you. Refamiliarize yourself with the metric system so that you can do such conversions effortlessly.

1.8 Graphing

Many experiments in this book will require you to make graphs as you analyze your data, so we will spend some time reviewing graphing basics. The first part assumes that you will be drawing the graph by hand instead of using a computer program. Understanding these concepts before trying to make a computer do this for you is important.

Drawing Graphs by Hand

A graph is a plot of two quantities. One is something that you control and is called the **independent variable;** usually it goes on the x axis. The other is the quantity that changes as you change the independent variable. It is called the **dependent variable** and goes on the y axis. For instance, if you put varying quantities of protein in a series of test tubes and then measure the absorbance, the variable you control is the quantity of protein, so it goes on the x axis. The absorbance is what changes as you change the protein quantity, so it goes on the y axis. By convention, "plot p versus q" means that p is on the y axis. **Always use graph paper!** A graph scribbled on binder paper is *not* a graph.

Units and Exponents Sometimes it is challenging to figure out what values to plot. Let's say we did a protein determination and recorded the adjacent data.

Quantity of Protein	Absorbance
0.001 mg	0.05
0.003 mg	0.16
0.005 mg	0.225
0.007 mg	0.33
0.008 mg	0.375

It would be particularly messy to plot those values in that form due to the number of zeros. The best thing to do is to change the units so that easier numbers can be plotted. The easiest way is to plot micrograms (μg) instead of milligrams (mg). That makes nice numbers like 1, 3, 5, 7, and 8 (Figure 1.1). Another way is to use exponents. Most science students make a mistake with the usage of exponents because most graphs that they have seen have not been done with standard scientific nomenclature. Notice that the x axis is labeled mg $\cdot$ 10^3, *not* $\cdot$ 10^{-3}. For scientific writing, the exponent indicates what you have already done to the number to put it on the graph with that scale, not what you expect the reader to do. In other words, the first value was 0.001 mg. To get rid of all of the zeros, we multiplied by 1000 to give a final value of 1. Therefore, on the axis we indicate that the value represents the milligrams of protein multiplied by 10^3.

Drawing the Line To make a graph, you must know something about the system and what you expect to see. When we do a protein assay, for example, we expect to get a straight line, at least to a certain limit of protein in the tube.

FIGURE 1.1 *A proper linear graph*

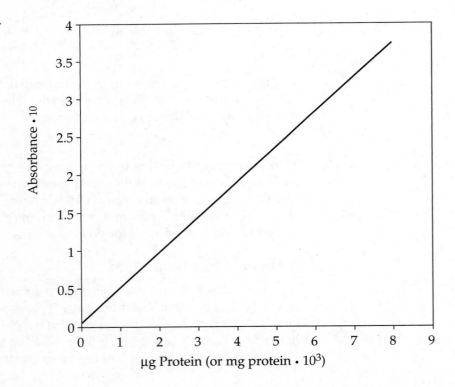

Therefore, we try to draw the best straight line that we can through the points. You can do this by using a computer program to draw the line for you (discussed later), by using a calculator with a statistics function to tell you the slope of the line, or by visually putting down the line that goes through the most points. In Figure 1.1, it is a mistake to just "connect the dots," thereby giving you a zigzaggy line.

When the relationship is expected to be linear, the best line is determined by **linear regression.** This is a mathematical process in which the line is placed to minimize the area between the points and the line. This is also called the **least-squares method.** Calculators and computers can do this easily, but your eyes cannot.

Constant Scale It is also important to remember that the scale must be constant. If you define 1 cm on the graph to be 1 μg as in Figure 1.1, then that scale must be maintained. A common mistake is to change the scale to fit your data instead of plotting your data on a given scale. This mistake is shown in Figure 1.2. This gives a very ugly line because the proper distance between the points is not maintained.

Use the Whole Graph Figures 1.1 and 1.2 are actually a bit small. It is generally best to use as much of the graph as possible. Also, the closer the line is to a 45° angle, the more accurate you will be when interpolating an unknown from it.

FIGURE 1.2 *An improper graph*

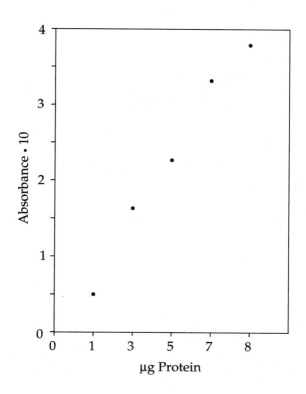

Must Zero Be a Point on the Graph? Most people naturally put the origin (0, 0) on the graph. Most of the time, that is correct, but there are times when it is inappropriate. A common reason not to include zero is when you are graphing log molecular weight (MW) versus mobility. This is done after doing an experiment with gel electrophoresis or gel filtration. For example, let's plot the log of MW versus mobility for the adjacent values.

If we start the graph at the origin, we get a graph that looks like Figure 1.3. This graph is very compressed on the y axis and does not use much of the graph paper. There would be huge error if we tried to interpolate within this graph. For this type of graph, it is much better to start the graph at 4 on the y axis and go until 5, as in Figure 1.4.

MW	Log MW	Mobility (cm)
15,000	4.18	6
40,000	4.60	2.9
66,000	4.82	1
90,000	4.95	0.1

Log Scales You will make many graphs using a log scale such as in Figure 1.4. However, there is a much easier way to plot such data. Buy some semilog paper, which does the calculations for you. Log paper is divided into cycles. A cycle is for data points that are all within one power of 10. The data from the preceding example fit onto one cycle because 15,000 − 90,000 are all within one power of 10. If we want to plot the log of 200,000, then we would need a second cycle. With log paper, just plot the numbers 15,000, 40,000, 66,000, 90,000, and 200,000 directly on the graph on the y axis and the paper takes the log for you. **Do not take the log and try to plot 4.18, 4.60, . . . on log paper.** Figure 1.5 shows how these data are plotted on two-cycle semilog paper.

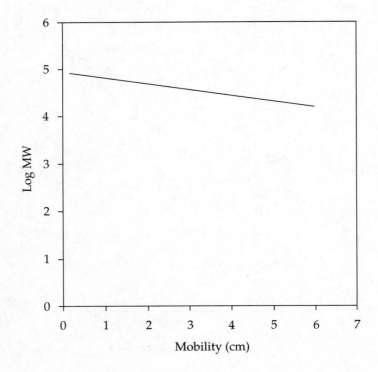

FIGURE 1.3 *An unacceptable log-scale graph*

FIGURE 1.4 *A proper log-scale graph*

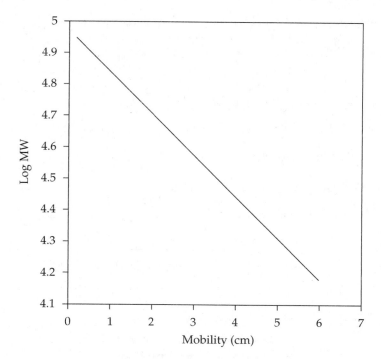

FIGURE 1.5 *Two-cycle semi-log graph*

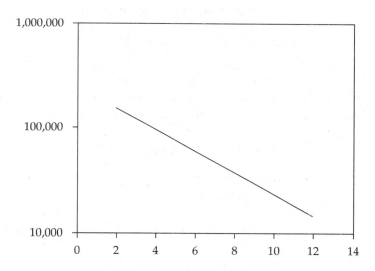

TABLE 1.3 *Absorbance versus Micrograms of Protein*

µg Protein	Absorbance
0	0
10	0.10
20	0.20
30	0.05
40	0.40
50	0.50

Bad Data Points The one thing that you can do manually better than a computer is decide when you have a bad data point. Let's say we want to plot the data for absorbance versus micrograms of protein shown in Table 1.3. If you plot the points on a standard graph, you will most likely notice that five of the six points fall perfectly on a straight line. However, one point (30, 0.05) is way off. Most computer programs default values would draw the graph shown in Figure 1.6 for these data. As you can see, the line is well below all data points except the bad one. Most of us would choose to

FIGURE 1.6 *Linear regression for a graph with a bad data point*

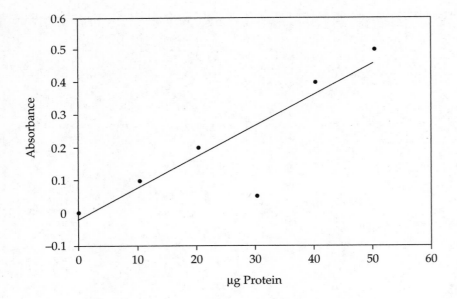

redo the 30-μg tube rather than report data of this caliber. If that were not possible, we would probably draw the line through the five perfect points and ignore the bad point altogether.

Graphing with Computers

Many of you probably have your own computer with a graphing program or spreadsheet program with graphing capability. We live in the computer age, and more classes are becoming computer driven. Many graphing programs are available; Sigmaplot, Kaleidagraph, Quattropro, and Excel are just a few. We use Excel as an example of how you can analyze data using spreadsheets and graphing programs.

Loading the spreadsheet The first step is to set up the spreadsheet. With Excel, you have a basic grid of columns, which are labeled A–Z, and rows, which are numbered 1 to several hundred. Each combination of a number and letter is called a **cell**. You start by putting data into the cells. Cells can hold data in the form of numbers or letters. For most applications that we are interested in, we put in numbers. If you want to plot absorbance versus micrograms of protein for the Bradford protein assay, you might have data that look like that shown in Table 1.4.

TABLE 1.4 *Absorbance versus Micrograms of Protein for Bradford Assay*

μg Protein	Absorbance
0	0
10	0.10
20	0.21
30	0.29
40	0.40
50	0.52

When you load this into Excel, you can put the column labels (μg Protein and Absorbance) into the spreadsheet, but that is not necessary when planning to use the data to make a graph. The program will "talk" you through creation of the graph, including labeling the axes. It is best to put the values that will be on the *x* axis into the left-hand column because it simplifies the graphing process. Figure 1.7 shows what the spreadsheet looks like.

Creating the Graph Once you have loaded the data into the spreadsheet, you are ready to create the graph. Excel provides a user-friendly process

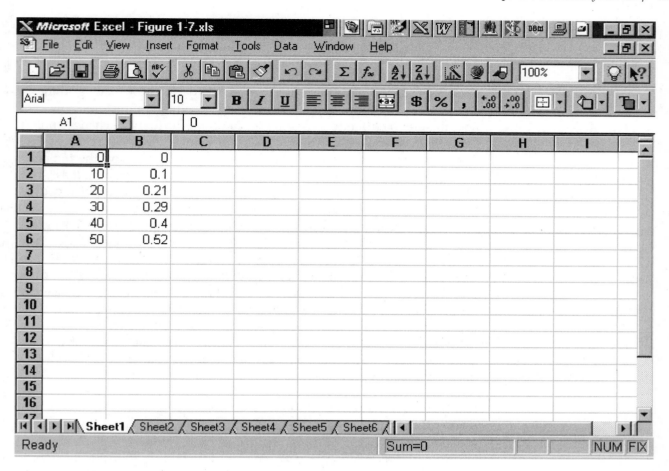

FIGURE 1.7 *Graphing data loaded into Excel spreadsheet*

involving dialog boxes. First, select the data that will be put into the graph by clicking on the A1 cell and dragging down and over until all data are highlighted. Then click on the Chart Wizard button. The first dialog box that pops up gives you a choice of chart types, such as bar graph, pie graph, and scatter plot. Once you choose a basic type, it gives you subchoices and pictures to help you choose. We will choose to do the XY scatter plot, which is normally what you will want to do. *Here is a potential danger point of using a computer to make a graph:* The default value for many computer programs has straight lines connecting the points (the dreaded connect-the-dot graph). This is usually wrong in biochemistry. What you want is a best-fit line of some type; in our case, it is a linear regression line. At this point in the process, choose the XY scatter plot that does not attempt to put any line over the points; the line will be put in later. With whatever program you are using, you must learn how to make the program give you the type of graph you want. Don't rely on the default values. With Excel, click on Next to continue the dialog. The next box that pops up (box two of four) asks about chart source data. If you have correctly selected the data from the columns you want, go on to the next box by clicking on Next. This brings us to box three of four in the dialog process. This is the chart-formatting step in which you decide how the graph is to look. Select the Title tab and then add in the graph title and the title of both axes. Select the Gridline tab that allows you to decide how many gridlines, if any, will appear on the graph. At each step, a minigraph picture shows you how the graph currently looks, so you can continue to make changes until it looks the way you want. Anytime that you change your mind, click on Back to go back to the last box. The fourth dialog box gives you the option of putting the graph as an object on the spreadsheet itself or putting it as a new sheet. Select the latter because it gives a better formatted page, which you can then print easily. The graph now looks like that in Figure 1.8.

To avoid creating the default connect-the-dots graph, we selected a graph that had no line; we still have to create the line. This is the one place in Excel that is not user friendly. You actually have to know something about the program, or you have to use the Help menu. In this case, we create the line using pull-down menus on the toolbar. Choose Chart and then Add Trendline from the pull-down menu. The box that pops up gives a picture of possible types of lines, including regression and polynomial. By choosing regression, we get the type of line we want for this type of linear data. An Options tab also lets us choose to put the regression coefficient and the mathematical equation of the line on the graph. We end up with the graph shown in Figure 1.9.

Nonlinear Graphs What do you do if the graph does not depict a linear relationship? Making such a graph starts out the same way, but then you use the Chart Wizard differently to customize your graph. A common example of this is analyzing an enzyme kinetics experiment, such as in Chapter 8. When enzyme velocity is plotted versus substrate concentration, a hyperbolic relationship is seen. Figure 1.10 shows typical enzyme

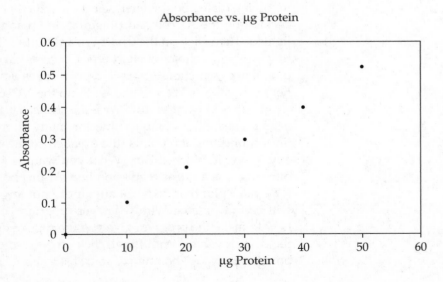

FIGURE 1.8 *Absorbance versus μg protein from the Excel graphing demonstration*

FIGURE 1.9 *Customized graph of data presented in Figure 1.7*

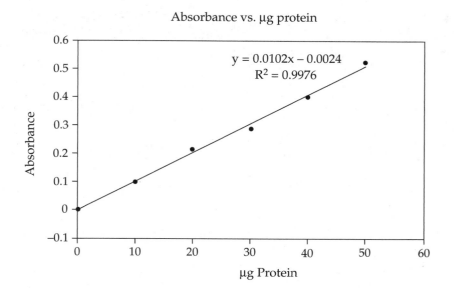

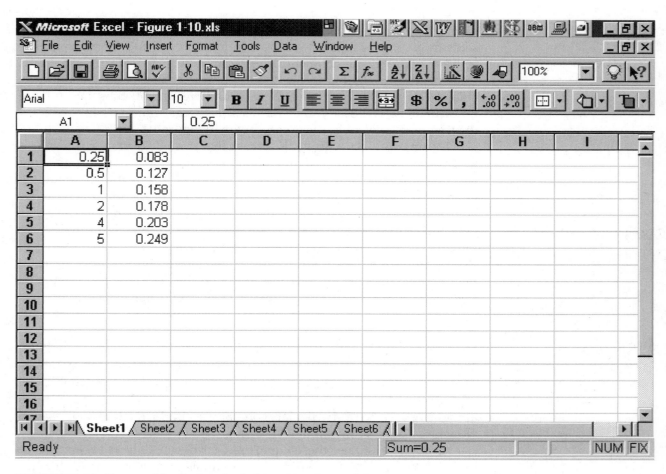

FIGURE 1.10 *Enzyme kinetic data loaded into Excel spreadsheet*

FIGURE 1.11 *Logarithmic graph of enzyme kinetic data (incorrect)*

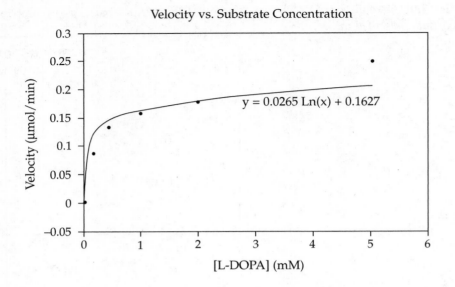

kinetic data as they appear in an Excel spreadsheet. The left column represents the concentrations of substrate in millimolars, and the right column is the enzyme velocity in micromoles per minute.

Once the original graph, without a line, is created, add a trendline as before. For hyperbolic graphs, there is no correct menu choice with Excel, unfortunately. The best you can do is select Logarithmic when you get to the trendline. This gives the graph shown in Figure 1.11.

This graph looks reasonable, except that it is a logarithmic function. A logarithmic graph is curved and similar to a hyperbolic one, but the equations do not match. This graph gives incorrect values if you attempt to interpolate kinetic constants from them. The good news is that you can use your spreadsheet to manipulate the data. If you take the reciprocals of the data and plot those, you plot a Lineweaver–Burk graph. Figure 1.12 shows the Excel spreadsheet with the reciprocal transformation.

Figure 1.13 shows the completed Lineweaver–Burk graph. By using the trendline options, it is possible to extend the line until it intercepts the *x* axis. This is necessary to measure the kinetic constants from this graph.

1.9 Pipets and Pipetmen

Much of your success in the biochemistry lab revolves around your ability to choose and use various devices for measuring and dispensing solutions. These range from simple glass tubes, called pipets, to highly advanced and expensive units collectively referred to as Pipetmen, although this is really a name given to a particular company's product, similar to saying "Kleenex" to mean any kind of cleansing tissue.

As you learn to use these liquid-transfer devices, always keep in mind that you are striving for two things—accuracy and precision. **Accuracy** is the relation between the volume you dispense and the volume you wanted

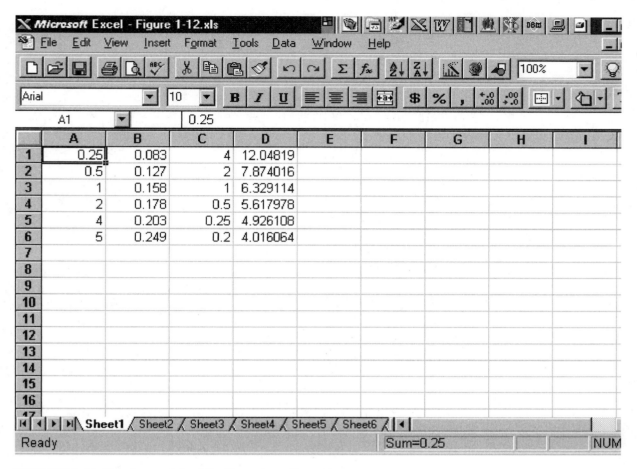

FIGURE 1.12 *Excel spreadsheet for linear transformation*

FIGURE 1.13 *Reciprocal graph of enzyme kinetic data*

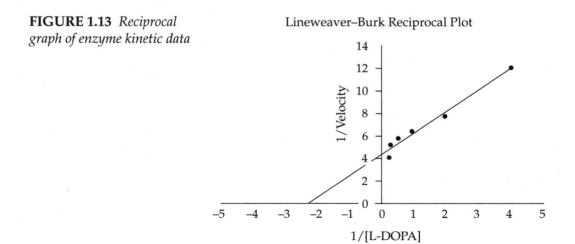

Lineweaver–Burk Reciprocal Plot

FIGURE 1.14 *Volumetric pipet*

to dispense. If you have a pipet and are trying to dispense 100 μL but actually dispense 70 μL, the volume was very inaccurate. **Precision** has to do with reproducibility. If you are to dispense 100 μL into each of five tubes and what you actually dispense is 70.0 μL, 69.9 μL, 70.2 μL, 70.1 μL, and 69.8 μL; then the volumes were very inaccurate but were very precise. Many of the mechanized Pipetmen-type dispensers are famous for their precision, but their accuracy depends greatly on *your* technique and how often they are calibrated.

Volumetric Pipets

A volumetric pipet is calibrated to deliver one set volume. An example is shown in Figure 1.14. The tolerance for error is sometimes written on the pipet. For example, a 10-mL pipet may say ±0.2%, and that gives the expected accuracy of the pipet.

Measuring Pipets

These pipets are graduated to indicate varying volumes of liquid, so you can use them to deliver many different volumes up to the maximum volume of the pipet. There are two basic types. **Serological pipets,** often called **blowout pipets,** are graduated to include the volume all the way to the tip (Figure 1.15). If you have a 10-mL serological pipet and you want to deliver 10 mL, bring the volume all the way up to the zero line and then expel all the liquid. **Mohr pipets** are not graduated to the tip; to dispense the same 10 mL, let the meniscus run down from the 0 line and then stop it at the 10 line (Figure 1.16).

Because your accuracy depends on how well you can get the meniscus to start and stop on the lines you want, a pipet is most accurate when used closest to its full capacity. A simple calculation will verify this. Suppose you are going to dispense 10 mL using a Mohr pipet that has graduations of 0.1 mL. If your hand–eye coordination is such that you routinely have an error of ±0.1 mL, you might accidentally stop the meniscus on the 9.9 line. That is an error of 0.1 mL/9.9 mL = 0.01, or 1%. Now, if you try to use the same 10-mL pipet to dispense 1 mL, draw the solution up to the 0 line and attempt to stop it at the 1-mL mark. If you have the same inaccuracy in your technique, you actually stop it at the 0.9-mL mark. The error is 0.1 mL/0.9 mL = 0.11, or

FIGURE 1.15 *Serological, or blowout, pipet*

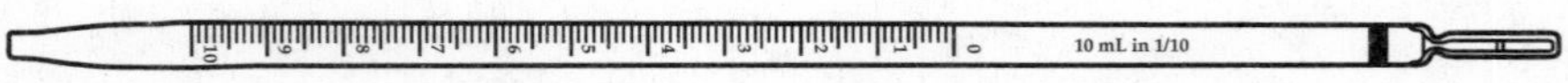

FIGURE 1.16 *Mohr pipet*

TIP 1.4 Prewetting the pipet tip can help if you are pipetting liquids with a low surface tension, such as organics. They tend to drip out of the pipet tip the first time you load them. Draw up the organic, then blow it out. Draw up some more, and this time it should hold without dripping.

11%. That is why you should always choose the pipet that you can fill the most. It makes sense to use that 10-mL pipet to pipet anything from 5.1 to 10 mL, but if you want to pipet only 4 mL, then you should choose the common 5-mL pipet. If you want to pipet 2 mL, then the 5-mL pipet would not be so good because there is also a common 2-mL pipet. Make sure you know what pipets are available and use the ones that can be used near to capacity.

To use glass pipets correctly, follow this procedure:

1. Using a pipet pump or rubber bulb, draw the solution into the pipet up past the starting mark.

2. Depending on the device you use to draw up the solution, either leave the device attached or remove it and use your index finger to control the flow.

3. Wipe off the tip with a tissue.

4. Touch the tip of the pipet to the sides of the vessel you took the solution out of and let it run down until the meniscus is at the starting line.

5. Transfer the pipet to the vessel you want to put the solution into and touch it to the sides of the vessel.

6. Let the solution run down the sides until the meniscus is at the stopping line.

Glass Syringes

One of the most accurate devices is the glass syringe, the most common of which is the Hamilton syringe (Figure 1.17). These have a tight-fitting metal plunger in a glass tube and often are used for accurate dispensing of volumes as small as 1 μL.

To use a Hamilton syringe correctly, follow this procedure:

1. Draw the solution into the syringe and expel several times to wet the inside of the glass and the needle.

FIGURE 1.17 *The Hamilton syringe*
(Courtesy of Hamilton Co.)

2. Slowly draw up the solution until the plunger is at the line you want.

3. Watch carefully for bubbles forming. If you get bubbles between the solution and the plunger, expel the solution quickly to shoot out the bubble.

4. Repeat until you have no bubbles in the syringe.

5. Wash the syringe out repeatedly with water. Never leave salts or organics in the syringe, or the plunger will become permanently locked.

6. Be careful not to bend the plunger. They bend easily and once bent are useless.

Pipetmen

The generic term for these is air-displacement piston pipets, so you can see why we say Pipetmen. The three most common types in use these days are the Pipetman by Rainin Instruments (Figure 1.18), the Eppendorf pipettor

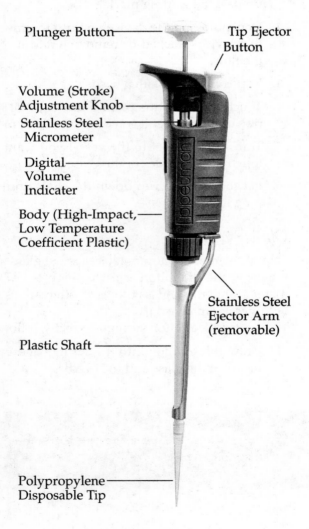

FIGURE 1.18 *Rainin Pipetman*
(Courtesy of Rainin Instruments, Inc.)

Plunger Button

Tip Ejector Button

Volume (Stroke) Adjustment Knob

Stainless Steel Micrometer

Digital Volume Indicater

Body (High-Impact, Low Temperature Coefficient Plastic)

Stainless Steel Ejector Arm (removable)

Plastic Shaft

Polypropylene Disposable Tip

TABLE 1.5 *Volume Ranges of Rainin Pipetmen*

Pipetman Model	Range (μL)		Smallest Increment (μL)
	Adjustable	Recommended	
P-2	0–2	0.1–2	0.002
P-10	0–10	0.5–10	0.02
P-20	0–20	2–20	0.02
P-100	0–100	10–100	0.2
P-200	0–200	50–200	0.2
P-1000	0–1000	100–1000	2.0
P-5000	0–5000	500–5000	2.0

by Brinkman Instruments, and the Integrapette by Integrated Instrument Services. We use the Pipetman as an example throughout this section.

Pipetmen come in sizes from the P-2, with a maximum volume of $2\,\mu L$, to the P-5000, with a maximum volume of $5\,mL$. The most common sizes are the P-20, P-200, and P-1000. The instrument uses disposable plastic tips in various sizes depending on the size of the Pipetman. There is a color-coding system between the pipet and the tips used. For example, the P-1000 has a blue top to indicate that it should use blue pipet tips, which are the size that goes up to $1\,mL$. The P-20 and P-200 both have yellow tops and use yellow tips, which go up to $200\,\mu L$. Before you start, make sure you have the proper size Pipetman and the proper tips to use.

Most Pipetmen are infinitely variable from a lower limit up to the maximum. Never try to dial the Pipetman past its maximum because it will irreversibly damage the unit. As you set the volume on the Pipetman, keep in mind what the maximum is. The volume dial has three digits, and what those three digits mean depends on the total volume possible for the pipet. For example, if the dial reads 050, that is $500\,\mu L$ for a P-1000 or $50\,\mu L$ for a P-200. Table 1.5 shows the recommended range for the Rainin Pipetman, as well as a description of the dial readings for each one. The other brands of pipettors are similar.

Accuracy of Pipetmen It is very easy to establish the accuracy of a pipettor. All you have to do is pipet a known volume of water onto a balance and weigh it. If you have a digital top-loader balance in the lab, this takes but a few seconds and could save you hours of frustration later. In a teaching lab, verifying that the instrument you are about to use is accurate is especially important. If you have a P-1000, $1000\,\mu L$ ($1\,mL$) should weigh $1\,g$. If you have a P-200, $200\,\mu L$ should weigh $0.2\,g$, and so on.

How to Use a Pipetman

1. Make sure you have the correct size Pipetman and tips.

2. Place the tip on the pipettor and tighten with your fingers, being careful not to touch the very tip. Wear gloves when appropriate.

3. Hold the pipettor vertically over the solution you want to draw up. Push the plunger down to the first stop. Be careful to not drive past the first stop.

4. Put the tip into the solution to a depth of a few millimeters.

5. Slowly let the plunger come up as the solution is drawn into the tip. After the plunger is all the way up, hold the tip in the solution for a couple of seconds to make sure that the solution is not still moving.

6. Look at the tip to make sure that you drew up the correct amount of solution.

7. Move the pipetman to the vessel that you want to dispense into. Touch the tip to the side of the vessel.

8. Slowly depress the plunger to the first stop and then past the first stop to the second stop. This will blow the solution out the tip. Any solution remaining should not be blown out further.

9. Use the tip ejector to dispose of the tip.

ESSENTIAL INFORMATION

Nearly all scientific calculations involve units. An answer without units is a wrong answer. We use units based on those of the MKS system. These units often have prefixes to tell you the power of 10 based on the standard unit. Learn how to convert from one prefix to another without needing to write long conversion factors. You will need to do this hundreds of times during a biochemistry course. Memorize the most used prefixes, such as kilo-, milli-, micro-, and nano-.

Your results will depend largely on how well you pipet. Learn how to use all the available glass pipets and automatic pipettors. You must know when to use a particular liquid-transfer device and when not to. Practice drawing up standard volumes into the Pipetmen. Then, during an experiment, you will know if you have pipetted the correct amount.

Study the section on graphing carefully. You cannot make a graph if you do not know what the relationship is between the x and y coordinates. Always use graph paper or a computer graphing program. Never scribble a graph on regular binder paper. You should rarely connect the dots on a graph. You should almost always draw a best-fit curve. Be wary of the default values on your computer program. Make the computer work for you, not the other way around.

Experiment 1

Using Pipettors

In this experiment, you will learn how to use Pipetmen of various sizes and measure their accuracy, precision, and calibration. The questions at the end will also review concentration and dilution.

Prelab Questions

1. What is the usable range of a P-1000 Rainin Pipetman?

2. What is the difference between *accuracy* and *precision*?

3. What should 100 μL of water weigh?

4. What should 1000 μL of water weigh?

Objectives

Upon successful completion of this lab, you should be able to

- Become familiar and adept at using pipettors.
- Check the calibration of the pipettors.

Experimental Procedures

Materials

100- or 200-μL pipettors

1000-μL pipettor

Yellow and blue pipet tips (or otherwise correct size)

Deionized water

Balances

Methods

Part A: Precision of P-100 or P-200 Pipettor

1. Acquire a P-100 or P-200 pipettor with the correct size tips. Make sure that the color matches.

2. Set the pipettor to $100\,\mu L$.

3. Place weighing paper or a weighing boat on the balance and tare the weight to zero.

4. Draw up the $100\,\mu L$ of deionized water and dispense it onto the weighing paper or weighing boat. Record the weight of the water.

5. Repeat the procedure twice more.

6. Draw up 10, 20, 50, and $75\,\mu L$ just to see what they look like in the tip.

Part B: Precision of P-1000 Pipettor

1. Acquire a P-1000 pipettor and the correct size tips. Make sure that the color matches.

2. Set the pipettor to $1000\,\mu L$.

3. Place weighing paper or a weighing boat on the balance and tare the weight to zero.

4. Draw up $1000\,\mu L$ of deionized water and dispense it onto the weighing paper or weighing boat. Record the weight of the water.

5. Repeat the procedure twice more.

6. Set the pipettor to $100\,\mu L$ and check the weight of the liquid three times.

7. Draw up 200, 500, and $750\,\mu L$ of water just to see what they look like in the tip.

8. Check the analysis of results questions, make sure that you have all the data you need, and put the pipettor away.

Name _____   *Section* _____

Lab partner(s) _____   *Date* _____

Analysis of Results

Experiment 1: **Using Pipettors**

Part A: Precision of P-100 or P-200 Pipettor

1. Record the weight you measured for the three trials of 100 µL:

   Weight 1 (x_1) _____

   Weight 2 (x_2) _____

   Weight 3 (x_3) _____

2. Average the three weights.

   Average of three trials: _____

3. Calculate the % error between the average of the three trials and the true value:

$$\% \text{ Error} = \frac{|\text{avg weight} - 0.100\,\text{g}|}{0.1\,\text{g}} \times 100 = \underline{\hspace{3cm}}$$

4. Calculate the mean deviation for the three trials:

$$\text{Mean deviation} = \frac{\Sigma |x_i - x_{avg}|}{3} = \underline{\hspace{3cm}}$$

Part B: Precision of P-1000 Pipettor

1. Record the weight you measured for the three trials of 1000 µL:

   Weight 1 (x_1) _____

   Weight 2 (x_2) _____

   Weight 3 (x_3) _____

2. Average the three weights.

   Average of three trials: _____

3. Calculate the % error between the average of the three trials and the true value:

$$\% \text{ Error} = \frac{|\text{avg weight} - 1.00\text{ g}|}{1.00\text{g}} \times 100 = \underline{\hspace{3cm}}$$

4. Calculate the mean deviation for the three trials:

$$\text{Mean deviation} = \frac{\Sigma |x_i - x_{\text{avg}}|}{3} = \underline{\hspace{3cm}}$$

5. Record the weight you measured for the three trials of 100 μL using the P-1000:

Weight 1 (x_1) _____

Weight 2 (x_2) _____

Weight 3 (x_3) _____

6. Average the three weights.

Average of three trials: _____

7. Calculate the % error between the average of the three trials and the true value:

$$\% \text{ Error} = \frac{|\text{avg weight} - 0.100\text{ g}|}{0.1\text{ g}} \times 100 = \underline{\hspace{3cm}}$$

8. Calculate the mean deviation for the three trials:

$$\text{Mean deviation} = \frac{\Sigma |x_i - x_{\text{avg}}|}{3} = \underline{\hspace{3cm}}$$

Part C: Pipettors in the Lab

1. Which of the two pipettors that you used was the more accurate? Explain.

2. Which of the two pipettors that you used was the more precise? Explain.

3. What are the take-home messages from this exercise? Give three specific things that you learned from this lab.

4. Without checking the accuracy of a given Pipetman, would you predict that it is better to use a P-200 or P-1000 to pipet 100 μL? Why?

5. Is a Pipetman more like a serological pipet or a Mohr pipet? Why?

6. If you are trying to pipet an unknown liquid with a Pipetman and the liquid keeps running out of the tip before you can transfer it, what are two possible reasons for this? What can you do to remedy the situation?

7. How do you make 200 mL of a 0.1-M solution of a substance that has a molecular weight of 121.1 g/mol?

8. If you take 10 mL of the solution you made in Question 7, add 90 mL of water, mix, and then take 5 mL of the mixture and bring it to 25 mL, what will be the concentration of the final solution in molars, millimolars, and micromolars?

Additional Problem Set

1. How many grams of solid NaOH are required to prepare 200-mL of a 0.05 M solution?

2. What will be the concentration from Problem 1 expressed in % w/v?

3. How many milliliters of 5 M NaCl are required to prepare 1500 mL of 0.002 M NaCl?

4. What will be the concentration of the diluted solution from Problem 3 expressed in millimolars, micromolars, and nanomolars?

5. A solution contains 15 g of $CaCl_2$ in a total volume of 190 mL. Express the concentration in terms of grams/liter, % w/v, molars, and millimolars.

6. Given stock solutions of glucose (1 M), aparagine (100 mM) and NaH_2PO_4 (50 mM), how much of each solution do you need to prepare 500 mL of a reagent that contains 0.05 M glucose, 10 mM asparagine, and 2 mM NaH_2PO_4?

7. Calculate the number of millimoles in 500 mg of each of the following amino acids: alanine (MW = 89), leucine (MW = 131), tryptophan (MW = 204), cysteine (MW = 121), and glutamic acid (MW = 147).

8. What molarity of HCl is needed so that 5 mL diluted to 300 mL will yield 0.2 M?

9. How much 0.2 M HCl can be made from 5.0 mL of 12.0 M HCl solution?

10. What weight of glucose is required to prepare 2 L of a 5% w/v solution?

11. How many milliliters of an 8.56% solution can be prepared from 42.8 g of sucrose?

12. How many milliliters of $CHCl_3$ are needed to prepare a 2.5% v/v solution in 500 mL of methanol?

13. If a 250-mL solution of ethanol in water is prepared with 4 mL of absolute ethanol, what will be the concentration of ethanol in % v/v?

Webconnections

For a list of websites related to the material covered in this chapter, go to **Webconnections** at the *Experiments in Biochemistry* site on the Brooks/Cole Publishing web page. You can access this page as http://www.brookscole.com. Webconnections are in the biochemistry portion of the chemistry page under the title of this manual.

References and Further Reading

Boyer, R. F. *Modern Experimental Biochemistry.* Menlo Park, CA: Addison-Wesley, 1993.

Cleveland, W. S. *The Elements of Graphing Data.* Belmont, CA: Wadsworth, 1985.

Jack, R. C. *Basic Biochemical Laboratory Procedures and Computing.* New York: Oxford University Press, 1995.

Rainin Instrument Co., Inc. "Instructions for Pipetman," Woburn, MA, 1991.

Robyt, J. F., and B. J. White. *Biochemical Techniques.* Long Grove, IL: Waveland Press, 1990.

Segel, I. H. *Biochemical Calculations.* New York: Wiley Interscience, 1976. Still the best source of practice problems in basic biochemistry.

Steel, R. G. D., and J. H. Torrie. *Principles and Procedures of Statistics: A Biometrical Approach,* 2nd ed. New York: McGraw-Hill, 1980.

Chapter 2

Acids, Bases, and Buffers

TOPICS

Introduction

All biochemical reactions occur under conditions of strict control over the concentration of hydrogen ion. Biological life cannot withstand large changes in hydrogen-ion concentration, which we measure as the pH. When chemicals that keep the pH from changing are present, we say that the system is buffered. Whether in your body or in a test tube, the reactions that will be important to you will be buffered. The proper choice and preparation of a buffer is paramount to your success in a biochemistry lab.

Before we can truly understand buffers, however, we must understand the more basic concepts of strong versus weak acids and pH. Throughout this text, we will use the Brønsted definition of acid *as a substance that can donate a hydrogen ion and a* base *as a substance that can accept a hydrogen ion.*

2.1 Strong Acids and Bases

An acid is a compound that has a hydrogen ion that it can give up to the solution. Common strong acids that you are familiar with are hydrochloric acid (HCl), sulfuric acid (H_2SO_4), and nitric acid (HNO_3). When strong acids are dissolved in water, they dissociate completely into their ions:

$$HCl(aq) \rightarrow H^+(aq) + Cl^-(aq)$$

A more formal way of writing this equation would indicate that the hydrogen ion is not really "hanging out" loose. There are no "naked" protons, as we say; rather, the hydrogen ion is attached to a water molecule to give a hydronium ion:

$$HCl + H_2O \rightarrow H_3O^+(aq) + Cl^-(aq)$$

Let's use the abbreviated formulas for simplicity's sake.

Strong bases, such as NaOH, KOH, and $Ca(OH)_2$ likewise dissociate completely into ions. For example,

$$NaOH(aq) \rightarrow Na^+(aq) + OH^-(aq)$$

The strength of an acid is determined by how much of the hydrogen ion dissociates when the acid is put into water. This can be determined from the K_a:

$$K_a = \frac{[H^+][A^-]}{[HA]}$$

The complete dissociation is indicated by a unidirectional arrow. If we calculate the K_a for the HCl, it would be much greater than 1. If we did the same thing for the NaOH (using the corresponding K_b), it would also be very large.

Because strong acids and strong bases break down completely into their ions, it is relatively easy to calculate the pH. As long as you know how many moles are in the starting compound, you will know how many ions you will get. To calculate the pH of a solution of a strong acid or strong base, we use the following procedures.

PRACTICE SESSION 2.1

What is the pH of 0.01 M HCl?

Because HCl dissociates completely, the concentration of the H^+ is also 0.01 M.

$$pH = -\log[H^+] = -\log 0.01 = 2$$
Answer pH = 2

PRACTICE SESSION 2.2

What is the pH of 0.01 M NaOH?

NaOH is a strong base, so it dissociates completely into Na^+ and OH^- ions. Therefore, the concentration of OH^- is also 0.01 M.

This is also easy because we know that the product of the concentration of hydrogen ion and the concentration of hydroxide ion is always equal to 10^{-14}. This is called the *water equation*.

$$[H^+][OH^-] = 10^{-14} \, M^2$$

$$[H^+] = \frac{10^{-14} \, M^2}{0.01 \, M} = 10^{-12} \, M$$

$$pH = -\log 10^{-12} = 12$$
Answer pH = 12

2.2 Weak Acids and Bases

Determining the pH of solutions of weak acids or bases is a little trickier. Because they do not dissociate completely, determining the $[H^+]$ is more difficult, and an equilibrium expression with K_a must be used. The K_a tells us the strength of the acid, so it can be used to calculate how much H^+ dissociates from the acid. We would soon tire of writing equilibrium constants such as 10^{-8}, 4.3×10^{-6}, . . . , so we have simplified matters by using pK_a's. Basically the "p" of anything is $-\log$ of that quantity, just as the pH is $-\log[H^+]$. Therefore,

$$pK_a = -\log K_a$$

So if $K_a = 10^{-12}$, $pK_a = 12$. Since all reactions can be written in two ways, remember that when you use K_a's or pK_a's, you *must* write the reaction as an acid dissociation.

How you calculate the pH depends on which chemical species you have in solution. There are three possibilities: **weak acid only, weak base only,** or **buffer.** We consider only the first two here.

Weak-Acid-Only Situations

If we have 0.002 mol of the weak acid, acetic acid ($pK_a = 4.76$), and we bring the volume up to 100 mL with pure water, we will have a solution of HAc at 0.02 M. Some of that HAc will break down via the following equation:

$$HAc \rightleftharpoons H^+ + Ac^-$$

How much will break down? We don't know because it is a weak acid and doesn't break down completely. The pK_a is a measure of acid strength and indirectly will tell us how much breaks down. There are shortcut formulas available to do these calculations:

$$pH = \frac{pK_a - \log[HA]}{2}$$

$$= \frac{4.76 - \log(0.02)}{2}$$

$$= 3.23$$

Remember, use this equation when the problem gives you the amount or concentration of a weak acid and you have no way of determining the amount of H^+ or A^-. This shortcut formula works for most reasonable concentrations of acids that you will see in biochemistry. The shortcut formulas are valid for concentrations up to about 0.1 M and for pK_a values as low as about 3. Once the acid becomes much stronger than that or the total concentration is higher, the shortcut formulas start to break down. (See Section 2.11 for more information on these.)

Weak-Base-Only Situations

What if we start out with sodium acetate, NaAc. This is the weak base formed from acetic acid. If we start with a weak base, some of it will react with water via the following equation:

$$NaAc + H_2O \rightleftharpoons Na^+ + HAc + OH^-$$

How much will react this way? Again, if this had been a strong base, the calculation would be simple; however, because it is a weak base, we are not sure to what extent the reaction occurs. The pK_a indirectly tells us this. There is another equation to use for this situation:

$$pH = \frac{pK_a + 14 + \log[A^-]}{2}$$

$$= \frac{4.76 + 14 + \log(0.02)}{2}$$

$$= 8.53$$

2.3 Polyprotic Acids

The use of polyprotic acids, those with more than one hydrogen that can dissociate, creates one other complication. Each dissociation of a hydrogen has its own dissociation constant. If you look at Table 2.2, you will see several weak acids with more than one pK_a listed. This means they are polyprotic acids. Not to worry! The beauty of the system is that you normally have to worry about only one of them at a time. For example, if you start out with H_3PO_4, you have a weak-acid-only situation, and you use 2.12 for the pK_a. If you start out with Na_3PO_4, you have a weak-base-only situation, and you use 12.32 for the pK_a.

There is unfortunately one other possibility, and that is an **intermediate of a polyprotic acid.** If you put 0.01 mol of NaH_2PO_4 into water, what type of solution do you have? The active species is $H_2PO_4^-$, but what is it? It could be the acid in

$$H_2PO_4^- \rightleftharpoons H^+ + HPO_4^{2-}$$

or it could be the base in

$$H_2PO_4^- + H_2O \rightleftharpoons H_3PO_4 + OH^-$$

As it turns out, when you have an intermediate of a polyprotic acid, the pH is controlled solely by the two pK_a values and is independent of the concentration of the solution. The pH is calculated by using the following equation:

$$pH = \frac{pK_a1 + pK_a2}{2}$$

2.4 Buffers

What happens if we add both HAc and NaAc to the same solution? This is the definition of a buffer because it has both the weak acid and the weak base in the same solution. Some of the HAc would tend to break down via the equation

$$HAc \rightleftharpoons H^+ + Ac^-$$

and some of the NaAc would tend to break down via the other equation:

$$NaAc + H_2O \rightleftharpoons Na^+ + HAc + OH^-$$

This would be very confusing except that with weak acid systems these two competing reactions tend to cancel each other out and these reactions do not happen to any great extent because they are so weak.

The *Henderson–Hasselbalch equation* can be used to calculate the pH when you have a buffer:

$$pH = pK_a + \log\left(\frac{A^-}{HA}\right)$$

So, if we add 0.001 mol of HAc to 0.002 mol of NaAc and bring the volume up to 100 mL with pure water, the pH will be

$$pH = 4.76 + \log\left(\frac{0.002}{0.001}\right)$$
$$= 5.06$$

You may have noticed that, although we used three different equations for three different situations, we used the same pK_a each time. Why? The pK_a is really a number that is a constant for a reaction rather than a particular molecule. In each case, we dealt with acetic acid and acetate. The pK_a of 4.76 is the pK_a of the reaction

$$HAc \rightleftharpoons H^+ + Ac^-$$

Why a Buffer Is a Buffer

Buffers resist pH changes because they use up excess hydrogen ion or hydroxide ion. If we have a solution with both a weak acid and its salt and we add some H^+, then the following reaction occurs:

$$A^- + H^+ \rightarrow HA$$

TIP 2.1 A buffer is a solution that contains both the weak acid form and the weak base form. Your success with the write up for this chapter will depend on how well you recognize whether a solution is a weak acid, a weak base, or a buffer.

Conversely, if we add OH^-, the following occurs:

$$HA + OH^- \rightarrow A^- + H_2O$$

PRACTICE SESSION 2.3

Consider the following two systems:

System A: 0.1 L of pure water at pH 7

System B: 0.1 L of 0.1 M phosphate buffer at pH 7.2

If we now add 10 mL of 0.001 M HCl to each, what will be the pH in each solution?

System A: 10 mL = 0.01 L of 0.001 M HCl

$$0.001 \text{ mol/L} \times 0.01 \text{ L} = 1 \times 10^{-5} \text{ mol } H^+$$

Because the final volume is now 0.1 L + 0.01 L = 0.11 L, $[H^+]$ = 1×10^{-5} mol/0.11 L = 9.09×10^{-5} M H^+.

$$\text{pH} = -\log 9.09 \times 10^{-5} = 4.04$$

System B: At pH 7.2, which is the second pK_a for phosphoric acid, $HA = A^-$.

$$0.1 \text{ M phosphate buffer} \times 0.1 \text{ L} = 0.01 \text{ mol phosphate}$$

Because $HA = A^-$ and the total is 0.01 mol, there must be 0.005 mol of each form of phosphate, or 0.005 mol $H_2PO_4^-$ and 0.005 mol HPO_4^{2-}. Initially,

$$\text{pH} = pK_a + \log\left(\frac{[HPO_4^{2-}]}{[H_2PO_4]}\right)$$

$$7.2 = 7.2 \; + \log\left(\frac{0.005}{0.005}\right)$$

Now we add 1×10^{-5} mol H^+. For every mole of H^+ added, the following reaction occurs:

$$H^+ + HPO_4^{2-} \rightarrow H_2PO_4^-$$

Therefore, 1×10^{-5} mol $H_2PO_4^-$ will be created, and 1×10^{-5} mol HPO_4^{2-} will be lost. Now the Henderson–Hasselbalch equation looks like this:

$$\text{pH} = 7.2 + \log\left(\frac{4.99 \times 10^{-3}}{5.01 \times 10^{-3}}\right) = \textbf{7.198}$$

As you can see, the buffer changed from pH 7.2 to 7.198 with the added acid, while the unbuffered system plummeted to pH 4.04. Had

---

ESSENTIAL INFORMATION

A buffer is a solution of a weak acid and its conjugate base. It resists pH change when reasonable amounts of both forms are present. A buffer is best when used close to its pK_a. Buffers are made by taking a weak acid and titrating with a strong base until the pH is correct or taking a weak base and adding a strong acid until the pH is correct. A buffer can protect against pH changes from added hydrogen ion or hydroxide ion as long as there is sufficient basic form and acid form, respectively. As soon as you run out of one of the forms, you no longer have a buffer. To be a good buffer, the pH of the solution must be within 1 pH unit of the pK_a. The pK_a gives an idea of the strength of the acid and tells us what the pH will be when there are equal amounts of acid form and basic form present.

---

you been trying to run a reaction in water, your system's pH would have plummeted, and your experiment would have been ruined.

2.5 Good's Buffers

The original buffers used in the lab were made from simple weak acids and bases, such as acetic acid, phosphoric acid, and citric acid. It was eventually discovered that many of these buffers had limitations. They often changed their pH too much if the solution was diluted or if the temperature was changed. They often permeated cells in solution, thereby changing the chemistry of the interior of the cell. The scientist N. E. Good developed a series of buffers that are zwitterions, molecules with both a positive and negative charge. These do not readily permeate cell membranes and are more resistant to concentration and temperature changes.

Most of the common synthetic buffers used today have strange names, which you will quickly forget, and even stranger structures. Table 2.1 gives a few examples. You don't really need to know the structure to use a buffer correctly. The important considerations are the pK_a of the buffer and the concentration you want to have. The Henderson–Hasselbalch equation works just fine whether or not you know the structure of the compound in question.

PRACTICE SESSION 2.4

What is the pH of a solution if you mix 100 mL of 0.2 M HEPES in the acid form with 200 mL of 0.2 M HEPES in the basic form?

You have both the acid and basic forms, so you have a buffer and need to use the Henderson–Hasselbalch equation. You must first find the pK_a, which we know is 7.55. Then you must calculate the ratio of the base to acid. The formula calls for the concentration, but in this situation the ratio of the concentrations will be the same as the ratio of the moles, which will be the same as the ratio of the volumes because both solutions had the same starting concentration of 0.2. Thus, we can see that the ratio of base to acid is 2 to 1 because we added twice the volume of base:

TABLE 2.1 *Acid and Base Forms of Some Useful Biochemical Buffers*

Acid Form		Base Form	pK_a
TRIS—H$^+$ (protonated form) (HOCH$_2$)$_3$CNH$_3^+$	N—tris[hydroxymethyl]aminomethane (TRIS) ⇌	TRIS (free amine) (HOCH$_2$)$_3$CNH$_2$	8.3
$^-$TES—H$^+$ (zwitterionic form) (HOCH$_2$)$_3$CNH$_2$CH$_2$CH$_2$SO$_3^-$	N—tris[hydroxymethyl]methyl-2-aminoethane sulfonate (TES) ⇌	$^-$TES (anionic form) (HOCH$_2$)$_3$CNHCH$_2$CH$_2$SO$_3^-$	7.55
$^-$HEPES—H$^+$ (zwitterionic form) 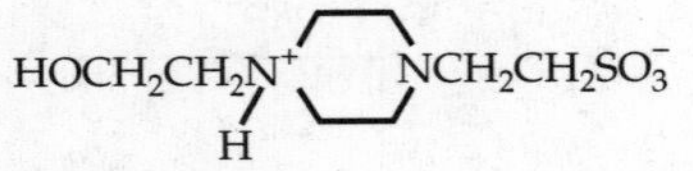 HOCH$_2$CH$_2$N$^+$H̶NCH$_2$CH$_2$SO$_3^-$	N—2—hydroxyethylpiperazine-N′-2-ethane sulfonate (HEPES) ⇌	$^-$HEPES (anionic form) 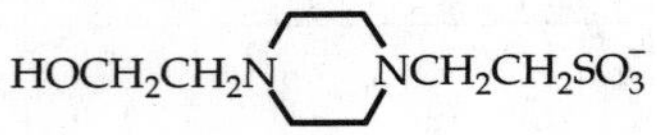 HOCH$_2$CH$_2$N̶NCH$_2$CH$_2$SO$_3^-$	7.55
$^-$MOPS—H$^+$ (zwitterionic form) 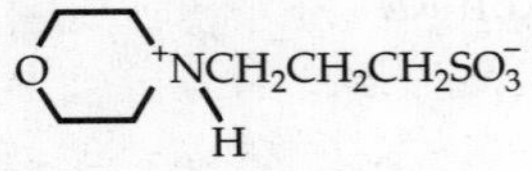 O̶$^+$NH̶CH$_2$CH$_2$CH$_2$SO$_3^-$	3—[N—morpholino]propane-sulfonic acid (MOPS) ⇌	$^-$MOPS (anionic form) 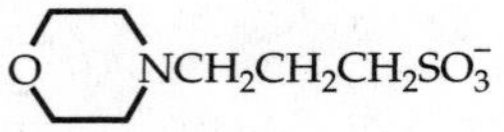 O̶N̶CH$_2$CH$_2$CH$_2$SO$_3^-$	7.2
$^{2-}$PIPES—H$^+$ (protonated dianion) 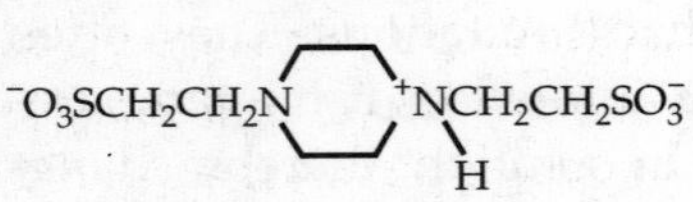 $^-$O$_3$SCH$_2$CH$_2$N̶$^+$NH̶CH$_2$CH$_2$SO$_3^-$	Piperazine—N,N′-bis[2-ethanesulfonic acid] (PIPES) ⇌	$^{2-}$PIPES (dianion) 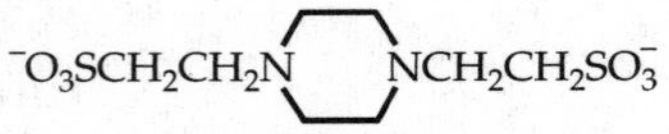 $^-$O$_3$SCH$_2$CH$_2$N̶NCH$_2$CH$_2$SO$_3^-$	6.8

$$pH = pK_a + \log\left(\frac{[A^-]}{[HA]}\right) = 7.55 + \log(2) = \mathbf{7.85}$$

2.6 Choosing a Buffer

When choosing a buffer, keep in mind the following:

1. You need both HA and A$^-$ to have a buffer. As soon as you run out of one of them, you no longer have a buffer.

   Think of a buffer system as a circle connecting the weak acid form and the weak base form (Figure 2.1). If you add H$^+$ to the system, the system shifts to produce more of the weak acid form. If you add OH$^-$

FIGURE 2.1 *The circle of buffers*

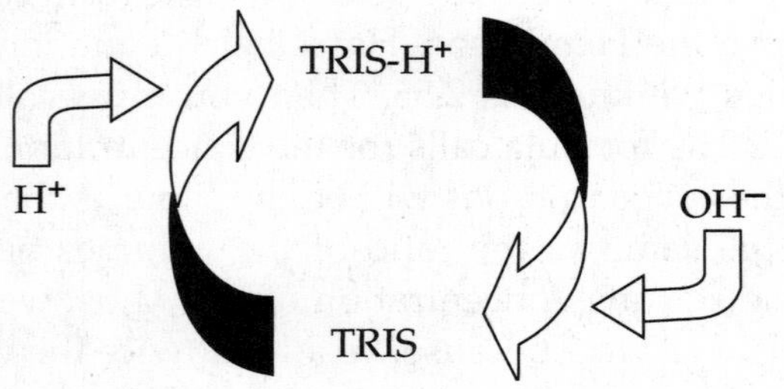

to the system, the system shifts to produce more of the weak base form. Either way you eliminate the excess H^+ or OH^-.

However, if you keep adding H^+, you eventually run out of the A^- form and no longer have a buffer. Therefore, the overall concentration of the buffer is important. If you have a buffer that is 0.1 M, you can add a lot more acid or base to it before it is exhausted than you can if you have a 0.001 M buffer.

2. The maximum buffering capacity is nearest the pK_a of the buffer. Can you see why? At a pH equal to the pK_a, equal amounts of the acid form and the basic form of the buffer are present. This is the best generic buffer that is equally good at buffering against added acid or added base. Therefore, if you are trying to choose the best buffer, choose the one with the pK_a closest to the pH you want.

3. There is a usable range for a buffer. Given that you may not be able to get a buffer with a pK_a equal to the pH you need, you may have to settle for less than perfect. There are limits to how far you can stray from the pK_a, however. How far away can you go and still have a buffer? If your ratio of HA to A^- is 10, then effectively 91% of the molecule is in the acid form, and 9% is in the basic form. This happens when your pH is 1 unit lower than the pK_a. At that point, not much of the base is left, so you cannot buffer very well against added acid. Conversely, if your ratio of HA to A^- is 0.1, then 91% of the molecule is in the basic form and, only 9% is in the acid form. That means that not much acid is left to buffer added base. This happens when the pH is 1 unit higher than the pK_a. By convention, we say that a buffer is only effective at its $pK_a \pm 1$ pH unit, but the closer the pH is to the pK_a, the better.

Table 2.2 gives some of the common buffers that you are likely to see, along with their pK_a's.

2.7 Effect of Concentration and Temperature

Effect of Temperature on Buffers

Buffers are at their best when the pH is adjusted correctly at the temperature at which they will be used. Most buffers are affected by temperature, some more than others. Table 2.2 shows the change in pK_a with temperature for some buffers. What this means is that if you calculate the change in pK_a with temperature, the pH would change the same amount because changing the temperature will not change the ratio of the basic form to acid form. Notice the sign on the number! For example, raising the temperature of HEPES by 1°C lowers the pH by 0.014.

Effect of Concentration on Buffers

Buffers are at their best when the pH is adjusted at the working concentration. However, most researchers make up concentrated solutions to save time and space. These solutions are then diluted to the appropriate concentrations before use. You will often see bottles labeled 20x TAE. This means that the

TABLE 2.2 *Buffers and Their Properties*

Compound	MW	pKₐ at 20°C	ΔpKₐ/°C
ACES	182.2	6.90	
ADA, free acid	190.2	6.60	−0.011
ADA, sodium salt	212.2	6.60	−0.011
BES	213.2	7.17	−0.027
Bicine	163.2	8.35	−0.018
Boric Acid	61.8	9.24	−0.018
CAPS	221.3	10.4	−0.009
CHES	207.3	9.5	−0.009
Citric Acid	192.1	3.14	−0.009
	192.1	4.76	−0.009
	192.1	6.39	−0.009
Glycylglycine	132.1	8.4	−0.028
HEPES, free acid	238.3	7.55	−0.014
HEPES, sodium salt	260.3	7.55	−0.014
Imidazole	68.1	7.00	−0.014
MES, free acid	195.2	6.15	−0.011
MOPS, free acid	209.3	7.20	−0.006
PIPES, free acid	302.4	6.80	−0.009
Phosphoric Acid (K_2HPO_4)	174.2	2.12	−0.009
	174.2	7.21	−0.009
	174.2	12.32	−0.009
TES, free acid	229.3	7.50	−0.020
Tricine	179.2	8.15	−0.021
Triethanolamine	185.7	7.66	−0.021
TRIS (Trizma base)	121.1	8.30	−0.031
TRIS-HCl	157.6	8.30	−0.031

solution is TAE buffer (TRIS, Acetate, EDTA) that must be diluted 20 to 1 before use. We have only talked about concentrations so far, not about activities. pH is really a function of activity, *a*. The activity *a* is a function of the concentration M and the activity coefficient γ. Activity coefficients change with concentration. The more charged the species is, the greater the change will be with changing concentration. The more exact equation for a weak acid is

$$K_{HA} = \frac{\gamma_H + \gamma_A - [H^+][A^-]}{\gamma_{HA}[HA]}$$

TABLE 2.3 *Activity Coefficients at Different Concentrations*

Ion	0.001 M	0.01 M	0.1 M
H^+	0.98	0.93	0.86
OH^-	0.98	0.93	0.81
Acetate	0.98	0.93	0.82
$H_2PO_4^-$	0.98	0.93	0.74
HPO_4^{2-}	0.90	0.74	0.45
PO_4^{3-}	0.80	0.51	0.16

Therefore, the K_a and pK_a will change with concentration even though the ratio appears to stay the same. Table 2.3 shows some typical changes.

What all this means to the pH of a buffer is rather complicated. Some buffers will show a pH increase with dilution, whereas others will show a pH decrease. The zwitterionic buffers, such as HEPES, do not have as pronounced an effect compared to phosphoric acid or citric acid.

2.8 How We Make Buffers

We don't have to add both HA and A^- to make a buffer. The easier way to do it is to titrate the HA with a strong base or the A^- with a strong acid. If we start with a weak acid, such as acetic acid, and we add a strong base, such as NaOH, we will quantitatively force the production of A^-:

$$HAc + NaOH \rightarrow H_2O + Ac^- + Na^+$$

This is not like a weak acid situation where we have to use the weak acid equation to figure out the extent of the reaction. Strong bases will always win the tug-of-war against a weak acid. For every molecule of NaOH that's put into the solution, one molecule of HAc will be converted to Ac^-.

Using the Henderson–Hasselbalch equation to determine how much weak acid and salt to add to make a buffer is appropriate if we don't have a pH meter available. Normally, one is available making a buffer of known pH and concentration much easier. Once you have the correct ratio of HA to A^-, it doesn't really matter how you arrived at that state.

To make a phosphate buffer with a concentration of 0.1 M and a pH of 6, rather than calculating the ratios of $H_2PO_4^-$ and HPO_4^{2-} that should be added, just start out with an amount of $H_2PO_4^-$ and add sodium hydroxide until the pH is 6. Then, by definition, we have a buffer at the correct pH. The only trick is getting the concentration correct. In this example, if we want to end up with 100 mL of buffer, take 50 mL of 0.2 M NaH_2PO_4 and add NaOH until the pH is correct and then add water until the volume is 100 mL. We then have the 0.1 M buffer that we need, and the pH is 6.

TIP 2.2 When working acid–base–buffer problems, spend more time thinking and less time pushing
keys on your calculator. Figure out first what type of
compound you are starting with. Is it a weak acid or a
weak base? Is it already a buffer?

2.9 The Big Summary

Summary of pH Equations

Definition	$pH = -\log[H^+]$
Water equation	$K_w = 10^{-14} = [H^+][OH^-]$
Henderson–Hasselbalch	$pH = pK_a + \log \dfrac{[A]}{[HA]}$
Weak acid only	$pH = \dfrac{pK_a - \log[HA]}{2}$
Weak base only	$pH = \dfrac{pK_a + 14 + \log[base]}{2}$
Polyprotic intermediate	$pH = \dfrac{pK_a1 + pK_a2}{2}$

In practice, diluting will affect the pH of a polyprotic acid solution, even
though the concentration does not appear in our shortcut formula. For any
solution, the pH should always be checked and readjusted after making
the final dilution.

PRACTICE SESSION 2.5

One Final Example Problem

Take 2.38 g of HEPES in the acid form and bring the volume to 100 mL
with water. This is solution A. Then add 3 mL of 1 M NaOH to make solution B. Finally, add 7 mL of 1 M NaOH to make solution C. What are the
pH values for the three solutions?

First determine how many moles you are dealing with. In this case,
the MW of HEPES is 238.3 g/mol. Thus,

$$moles = \frac{2.38}{238.3} = 0.01 \text{ mol HEPES acid}$$

Solution A: Because this solution was dissolved to 100 mL, the solution
is 0.01 mol/0.1 L for a concentration of 0.1 M. This is a weak-acid-only situation, so you use the weak-acid-only equation:

$$pH = \frac{pK_a - \log[HA]}{2} = \frac{7.55 - \log(0.1)}{2} = \mathbf{4.28}$$

Solution B: Next, 3 mL of 1 M NaOH were added to the solution, and that is 0.003 mol of OH^-, which will react with the weak acid to create a weak base until you run out of one of them:

$$0.01\,\text{mol HA} + 0.003\,\text{mol OH}^- \rightarrow 0.003\,\text{mol A}^- + 0.007\,\text{mol HA}$$

The 0.007 mol HA is what is leftover after the base reacts. The 0.003 mol of A^- is what is formed when the hydroxide reacts. You now have some HA and some A^-, so you have a buffer and now use the Henderson–Hasselbalch equation:

$$\text{pH} = \text{p}K_a + \log\left(\frac{\text{A}^-}{\text{HA}}\right) = 7.55 + \log\left(\frac{0.003}{0.007}\right) = \textbf{7.18}$$

Solution C: Finally, 7 mL of 1 M NaOH were added; that is another 0.007 mol of OH^-. How handy! This happens to be the exact amount of weak acid you had leftover from the last part. Thus, you will now use up all the weak acid and have only 0.01 mol of the basic form of HEPES left. This is a weak-base-only situation:

$$\text{pH} = \frac{\text{p}K_a + 14 + \log[\text{A}]}{2} = \frac{7.55 + 14 + \log\left(\dfrac{0.01}{0.110}\right)}{2} = \textbf{10.25}$$

For this equation, the only trick is to remember that it calls for the concentration of the weak base. Notice that the volume has changed during the experiment. That is why we divided 0.01 mol by 0.11 L because the new volume is the original volume of 100 mL (0.1 L) *plus* the total volume of base added (10 mL, or 0.01 L).

This problem can appear in slightly different forms in which we start with the weak base and go toward the acid or we start with a buffer and work back. The principles are the same.

2.10 Why Is This Important?

If you work in the sciences, you will use buffers every day. Many people use them blindly and make mistakes with them. This is even true of graduate students and post docs. If you understand this lab, you will be able to

---

TIP 2.3 If you pull a bottle of Trizma base off the shelf and read the label, you will see that the $\text{p}K_a$ is 8.3. If you then take a certain amount of the compound and add it to water and do nothing else, the pH will most certainly *not* be 8.3. It will be very much higher than 8.3. You only get the pH to equal the $\text{p}K_a$ if you titrate the Trizma base with strong acid until you have added half of an equimolar amount of acid.

choose the correct buffer for the job and understand how it works. You will know whether it will work well at different temperatures and concentrations. You will know the fastest way to make your buffer and be able to choose the best starting compounds. Making buffers is far from glamorous, but you will do it so many times that understanding it will save you time and money.

2.11 Expanding the Topic

Converting Ratios to Percentages

Many students find it difficult to calculate the percentage of a buffer that is in the weak acid or weak base form. This is usually because they have difficulty converting the ratios to percentages. For example, if using the Henderson–Hasselbalch equation and calculating that the ratio of A^-/HA is 1, then you would have no difficulty saying that the buffer is 50% A^- and 50% HA. If there is a total of 0.02 mol of buffer, there is 0.01 mol in the A^- form and 0.01 mol in the HA form.

What if the ratio is not so easy as that? Let's say you use the Henderson–Hasselbalch equation and find that the ratio is 2. This means that for every HA there are 2 A^-. That gives you percentages of 66% for A^- and 33% for HA. The ratio 2 to 1, means that a total of three parts are in the system. The numerator is $\frac{2}{3}$, and the denominator is $\frac{1}{3}$. To get the percentage, divide the numerator or denominator by the total.

If you use the Henderson–Hasselbalch equation and the ratio is 4, then that means A^-/HA is 4 to 1. A total of five parts is in the system, so the A^- is $\frac{4}{5}$, which is 80%.

Even if the ratio comes out to be really weird, like 4051 to 1, you will still calculate the percentages in the same way. If the ratio of A^-/HA is 4051 to 1, then there is a total of 4052 parts. The percentage of the buffer in the A^- form is $^{4051}/_{4052}$, or 99.975%.

History of the Shortcut Formulas

To determine the exact pH of a solution of weak acid, it is necessary to do some algebra that includes a quadratic equation. If we have a weak acid, HA, it will dissociate in water to give H^+ and A^- in equal amounts:

$$HA \rightarrow H^+ + A^-$$

If the initial concentration of HA is 0.1 M, then after it dissociates the concentration will be reduced by the amount that has dissociated. If we say that an unknown amount of H^+ will dissociate and we call that amount x, then the new concentration of HA will be 0.1 M $- x$. Recall that the acid dissociation expression is

$$K_a = \frac{[H^+][A^-]}{[HA]}$$

If we designate the amount of H^+ to be x, then the amount of A^- must also be x because they are formed in equal amounts from the dissociation. The equation is then

$$K_a = \frac{x^2}{0.1 - x}$$

This is a quadratic equation. To determine x and therefore the $[H^+]$, we need to solve for x in the equation. Most calculators can do this easily nowadays. However, if we make one simplifying assumption, this equation becomes much easier. Let's assume that the amount x is small compared to the initial concentration of HA. When will this happen? When the acid is weak, which is precisely why we are using this equation in the first place.

If we assume that x is negligible when compared to [HA], then the equation simplifies to the following:

$$K_a = \frac{[H^+][A^-]}{[HA]} = \frac{x^2}{[HA]} = \frac{[H^+]^2}{[HA]}$$

Solving for $[H^+]$ gives us

$$[H^+] = \sqrt{K_a[HA]}$$

To obtain an expression involving pH, put the expression into logarithmic form:

$$\log[H^+] = \frac{1}{2}\log K_a[HA] = \frac{1}{2}(\log K_a + \log[HA])$$

$$= \frac{-\log K_a - \log[HA]}{2}$$

$$= \frac{pK_a - \log[HA]}{2} = pH$$

Similar derivations can be done for the weak base solution and the Henderson–Hasselbalch equation (see *Biochemical Calculations,* by I. H. Segel). Because this equation is based on an assumption, anytime the assumption is wrong, so is the calculation. The stronger the acid, the more the real pH will deviate from that calculated by this formula.

Experiment 2

Preparing of Buffers

In this experiment, you will learn how to choose and prepare a buffer. You will then see how the pH of the buffer responds to dilution and compare how buffered and unbuffered systems respond to addition of acids and bases.

Prelab Questions

1. Calculate the weight of the buffers you will use to make the buffers for Part A for all the buffer possibilities listed under procedures in Part A, step 1. In other words, how many grams do you need to make 100 mL of a 0.1 M buffer?

2. If we give you HEPES in the basic form and ask you to make a buffer of pH 8.0, will you have to add HCl or NaOH? Why? (With commercial buffers, there is always an acid form and a basic form that can be bought. It is not obvious from the name of the compound, so look to see if it is acid or basic. If HEPES is bought in the acid form, then write the equation

$$HEPES \rightarrow HEPES^- + H^+$$

Objectives

Upon successful completion of this lab, you will be able to

- Calculate the pH of solutions of strong acids or bases, weak acids or bases, buffers, and/or combinations of these.

- Explain how a buffer resists change in pH.

- Prepare an appropriate buffer for a given pH.

- Calculate the theoretical pH of the buffer after adding a known quantity of acid or base.

- Predict the effect of changing temperature and concentration on buffer pH.

- Standardize and operate a pH meter.

Experimental Procedures

Materials

Solid buffers

Standard buffers: pH 4, 7, and 10

pH meters

1M HCl and NaOH

Solutions of unknown pH

Methods

Part A: Preparation of Buffers

Make two buffers starting with solid material, which is the most common way to make buffers. You will be given a desired pH, and your task is to prepare 100 mL of two appropriate buffers at a concentration of 0.10 M. One of the buffers will be a phosphate or citrate buffer, and the other will be one of the others (not phosphate or citrate).

1. Using the following table, choose the most appropriate buffer compounds for your pHs. Proceed with steps 2–7 for both buffers.

Buffer	pK_a1	pK_a2	pK_a3	Formula Weight (g/mol)
Acetate	4.76			136.1
CAPS	10.4			221.3
Citrate	3.06	4.74	5.40	294.1
HEPES	7.55			238.3
Phosphate	2.12	7.21	12.32	142.0
Tricine	8.15			179.2
TRIS	8.3			121.1

2. Calculate the weight of the buffer you would need to make 100 mL of a 0.100 M solution. Weigh out the correct amount and dissolve in 50 mL water.

3. Standardize the pH meter at pH 4, 7, and 10. Set up the beaker with your buffer solution on a stirplate such that you can stir the solution and read the pH continuously. If you have no stirplate, just swirl the beaker often while adding acid or base.

4. Use 1 M NaOH or 1 M HCl to titrate to the desired pH. Add the acid or base a drop at a time. By doing this, you effectively change some of the acid form to the basic form or vice versa until the ratio is the correct one to give you the pH you want.

5. Add water until the volume is about 99 mL.

6. Recheck the pH to make sure it has not changed. If it has, correct it with NaOH or HCl. **Warning! You might want to use a lower concentration of NaOH or HCl.**

7. Bring the volume to 100 mL and save this solution for later.

Part B: Effect of Concentration on pH

For this part, you must have a digital pH meter.

1. Take 10 mL of each of your two buffers and dilute with deionized water to give a final concentration of 0.01 M. Save these solutions.

2. Take 10 mL of your diluted buffers from step 1 and dilute to a concentration of 0.001 M. Save these solutions.

3. Measure the pHs of the undiluted and diluted solutions.

Part C: pH Measurement of Other Solutions

Measure the pH of the following solutions:

Distilled water

Unknown ⎯⎯⎯⎯

Unknown ⎯⎯⎯⎯

Part D: Why a Buffer Is a Buffer

1. Put 50 mL of one of your original 0.1 M buffers in a beaker. If your buffer has a pH higher than its pK_a, add 0.5 mL of 1 M HCl to it. Record the new pH. If your buffer has a pH lower than its pK_a, add 0.5 mL of 1 M NaOH to it. Record the new pH.

2. Repeat step 1, but use 50 mL of water instead of your buffer. Add either acid or base, depending on what you did in step 1.

Name _____

Section _____

Lab partner(s) _____

Date _____

Analysis of Results

Experiment 2: **Buffers**

Data

Part A

Buffer 1: _____ Weight (g): _____ Original pH:* _____

Buffer 2: _____ Weight (g): _____ Original pH:* _____

Part B

Buffer 1   pH of 0.1 M _____   pH of 0.01 M _____   pH of 0.001 M _____

Buffer 2   pH of 0.1 M _____   pH of 0.01 M _____   pH of 0.001 M _____

Part C

Distilled water pH: _____

Unknown _____   pH: _____

Unknown _____   pH: _____

Part D

Buffer chosen: _____ pH: _____ pK_a: _____

Acid or base added: _____

pH after adding acid or base: _____

pH of 50 mL of water: _____

pH after adding acid or base: _____

* When buffer powder was added to water.

Calculations

1. This problem will be done for one of the two buffers you made. (Your lab partner should do these calculations for the other one.)

   a. What is the ratio of A^-/HA in your buffer after you adjusted its pH to the required value?

   b. How many micromoles of A^- and HA are present in the solution?

   c. If you now add 3 mL of 1 M NaOH, will you still have a valid buffer?

2. Calculate the theoretical pH of one of your buffers at 0°C. Assume that room temperature is 22°C. If none of your buffers is listed on the table of changing pK_a with temperature, do this problem for TRIS at pH 8.0.

3. What is the most efficient way to make up a HEPES buffer at pH 8.5? What starting compounds and reagents will you use?

4. When Dr. Farrell was a graduate student, he once made up a pH 8.0 sodium acetate buffer. Why would the casual observer to this buffering faux pas come to the conclusion that he had the intellectual agility of a small soap dish?

5. If you make up a solution of 50 mL of 0.1 M TRIS in the acid form, what will be the pH?]

6. If you add 2 mL of 1 M NaOH to the solution in step 5, what will be the pH?

7. If you make up a solution of 100 mL of 0.1 M HEPES in the basic form, what will be the pH?

8. If you add 3 mL of 1 M of HCl to the solution in step 7, what will be the pH?

9. What can you conclude about the effect of dilution on the pH of a buffer?

Additional Problem Set

1. Calculate the pH of a 0.1 M HCl solution.

2. Calculate the pH of a 0.1 M NaOH solution.

3. What is the concentration of $[H^+]$ in molars, millimolars, and micro-molars for a solution of pH 5?

4. If you mix 10 mL of a 0.1 M HCl solution with 8 mL of a 0.2 M NaOH solution, what will be the resulting pH?

5. If a weak acid, HA, is 3% dissociated in a 0.25 M solution, calculate the K_a and the pH of the solution.

6. What is the pH of a 0.05 M solution of TRIS acid ($pK_a = 8.3$)?

7. What is the pH of a 0.045 M solution of TRIS base?

8. If you mix 50 mL of 0.1 M TRIS acid with 60 mL of 0.2 M TRIS base, what will be the resulting pH?

9. If you add 1 mL of 1 M NaOH to the solution in 6 above, what will be the pH?

10. How many total milliliters of 1 M NaOH can you add to the solution in Problem 6 and still have a good buffer (that is, within 1 pH unit of the pK_a)?

11. If you are making 100 mL of a 0.1 M HEPES buffer starting from HEPES in the basic form, is it prudent to get 50 mL of 1 M HCl from the community reagent bottle to use for your titration?

12. An enzyme-catalyzed reaction is carried out in a 50-mL solution containing 0.1 M TRIS buffer. The pH of the reaction mixture at the start was 8.0. As a result of the reaction, 0.002 mol of H^+ were produced. What is the ratio of TRIS base to TRIS acid at the start of the experiment? What is the ratio at the end of the experiment? What is the final pH?

13. The pK_a of HEPES is 7.55 at 20°C, and its MW is 238.31. Calculate the amounts of HEPES in grams and of 1.0 M NaOH in milliliters that would be needed to make 300 mL of 0.2 M HEPES buffer at pH 7.2.

Webconnections

For a list of websites related to the material covered in this chapter, go to **Webconnections** at the *Experiments in Biochemistry* site on the Brooks/Cole Publishing website. You can access this page at http://www.brookscole.com and follow the links from the chemistry page.

References and Further Reading

Boyer, R. F. *Modern Experimental Biochemistry*. Menlo Park, CA: Addison-Wesley, 1993.

Campbell, M., and S. Farrell. *Biochemistry*, 4th ed. Pacific Grove, CA: Brooks/Cole, 2002.

Ciullo, P. A. *Bicarb: Buffers, Bubbles, Buscuits, and Earthburps.* Buffalo Grove, IL: Maradia Press, 1997.

Jack, R. C. *Basic Biochemical Laboratory Procedures and Computing.* New York: Oxford University Press, 1995.

Mohan, C. "Buffers: A Guide for the Preparation and Use of Buffers in Biological Systems," La Jolla, CA: Calbiochem Co., 1995.

Robyt, J. F., and B. J. White. *Biochemical Techniques.* Long Grove, IL: Waveland Press, 1990.

Segel, I. H. *Biochemical Calculations*, 2nd ed. New York: Wiley Interscience, 1976.

Stenesh, J. *Experimental Biochemistry.* Boston: Allyn and Bacon, 1984.

Chapter 3

Spectrophotometry

TOPICS

Introduction

In this chapter, we deal with the most often used theories and techniques found in a biochemistry lab, those of spectrophotometry. Almost every experiment that you do in a lab involves the use of a spectrophotometer in some way. Along with using a Pipetman, how you use a spectrophotometer will affect the results of your experiments. If you learn to use them well, your labs will run smoothly. As with any piece of equipment, the old adage, "Garbage in, garbage out" is very true with even the simplest spectrophotometer.

3.1 Absorption of Light

Almost all biochemical experiments eventually use spectrophotometry to measure the amount of a substance in solution. **Spectrophotometry** is the study of the interaction of electromagnetic radiation with molecules, atoms, or ions.

Light, or electromagnetic radiation, has a wave and particle nature. The **wavelength λ** of light is the distance between adjacent peaks in the wave. The **frequency v** is the number of waves passing a fixed point per unit of time (see Figure 3.1).

FIGURE 3.1 *Wave nature of light*

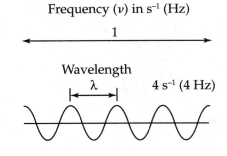

Frequency (v) in s⁻¹ (Hz)

1

Wavelength

λ

4 s⁻¹ (4 Hz)

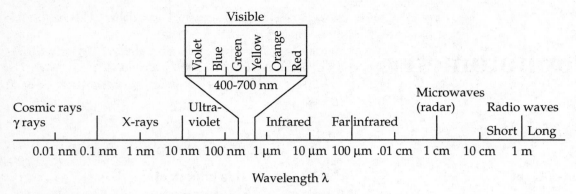

FIGURE 3.2 *Wavelength regions of light*

These parameters can be further defined by the equation

$$\lambda = \frac{c}{\nu}$$

where c is the speed of light.

Photons of different wavelength have different energies. These energies can be calculated by the equation

$$E = \frac{hc}{\lambda} = h\nu$$

where h is Planck's constant. Therefore, the longer the wavelength, the less energy the light has and vice versa.

Figure 3.2 shows the relationship between the wavelength of light and the common types of electromagnetic radiation. As you can see, those regions where the wavelength is very short correspond to the types of radiation that you know are powerful and often harmful, such as X-rays, gamma-rays, and ultraviolet radiation.

Most compounds have a certain characteristic wavelength or wavelengths of light that they absorb. Figure 3.3 diagrams this process. Thus,

FIGURE 3.3 *Absorption of light by a solution*

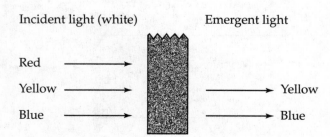

the solution looks green to us because green light (blue and yellow) is **transmitted** while the red light is **absorbed.**

A solution may contain many compounds that absorb at many different wavelengths. However, if a compound that we are interested in absorbs at a unique wavelength, we can determine its concentration even in a solution of other compounds.

3.2 The Beer–Lambert Law

If a ray of monochromatic light (one wavelength) of initial intensity I_0 passes through a solution, some of the light may be absorbed so that the transmitted light I is less than I_0 (Figure 3.4). The ratio of intensities I/I_0 is called the **transmittance** and is dependent on several factors:

1. If the concentration c of the absorbing solution increases, then the transmittance will decrease.

2. If the pathlength l that the light must travel through increases, then the transmittance will decrease.

3. If the nature of the substance changes or another substance that absorbs more strongly is used, then the transmittance will change. The nature of the substance is reflected in ε, the extinction coefficient, also called the absorptivity constant.

An equation can be written that incorporates these ideas:

$$\log \frac{I_0}{I} = \varepsilon l c$$

where $\quad I_0 \quad = \quad$ intensity of incident light

$\quad\quad\quad\quad I \quad = \quad$ intensity of transmitted light

$\quad\quad\quad\quad \varepsilon \quad = \quad$ extinction coefficient

$\quad\quad\quad\quad l \quad = \quad$ pathlength through solution

$\quad\quad\quad\quad c \quad = \quad$ concentration of absorbing solution

Log I_0/I is usually called the **absorbance** and is abbreviated A.

FIGURE 3.4 *Relationship between I, I$_0$, l, and c for a solution absorbing monochromatic light*

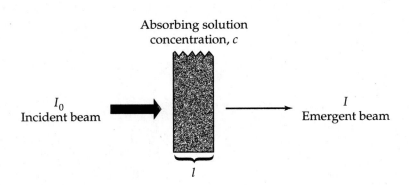

This law

$$A = \varepsilon l c$$

is called the Beer–Lambert law (Beer's law).

Some Points to Consider

1. Absorbance A has no units. It is just a number that can be read off of the spectrophotometer. The wavelength is often specified along with the absorbance, such as $A_{540} = 0.3$.

2. The extinction coefficient ε is the absorbance of a unit solution and has units of reciprocal concentration and pathlength. The most common ε values recorded are for a pathlength of 1 cm and a 1 M solution. Therefore, the expression $\varepsilon_{600} = 4000\,M^{-1}\,cm^{-1}$ means that a 1 M solution has an absorbance at 600 nm of 4000 if a 1-cm diameter cuvette is used.

3. Remember that the pathlength l is usually in centimeters and, if not specified, is assumed to be 1 cm.

4. The concentration c has units that are the reciprocal of the units for ε.

5. At least 2 mL of solution is needed in a cuvette in order to read it with the standard Milton Roy Spectronic 20 spectrophotometer.

6. Many things can interfere with your use of a spectrophotometer. If the cuvette is smudged or scratched, light will be scattered by the tube rather than absorbed by the solution. If you do not have sufficient volume (see point 5), the light may pass over the solution instead of going through it. The spectrophotometer must be well calibrated before use.

If a substance obeys the Beer–Lambert law, then a plot of A versus c is straight, as in the "ideal" line shown in Figure 3.5. More often, however, the line is curved, shown as "reality" in Figure 3.5.

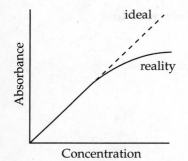

FIGURE 3.5 *Absorbance versus concentration*

PRACTICE SESSION 3.1

We measure the absorbance of a solution of compound X, which absorbs at 540 nm. The cuvette has a width of 1 cm, the extinction coefficient at 540 nm is 10,000 $M^{-1}\,cm^{-1}$, and the absorbance is 0.4. What is the concentration of compound X?

From Beer's law

$$A = \varepsilon l c$$

$$c = A/\varepsilon l = \frac{0.4}{10{,}000\ M^{-1}\,cm^{-1}}\ (1\ cm)$$

$$= 4 \times 10^{-5}\,M$$

ESSENTIAL INFORMATION

Beer's law enables you to calculate the concentration of a substance in solution after measuring the absorbance with a spectrophotometer. Before using a spectrophotometer, it must be properly calibrated, or zeroed. If it is not, the numbers generated will be meaningless.

As we will see in later chapters, you can also measure a changing concentration, such as the product of an enzymatic reaction, by measuring the changing absorbance. Beer's law only works if you know that the relationship between absorbance and concentration is linear. This is *not* always the case. If you cannot use Beer's law, proceed to the next technique, which is making a standard curve.

Reagent Blanks

A **reagent blank** is a control in which everything is included except the substance for which we are testing. One problem often encountered in spectrophotometry is that an absorbance is present at a given wavelength not due to the substance of interest. We handle that by mixing up all solutions in a tube except that substance and then read the absorbance. The absorbance of the reagent blank is then subtracted from the other readings.

PRACTICE SESSION 3.2

We want to read the absorbance at 595 nm of a protein solution mixed with the colorizing solution called Bradford reagent. What is a suitable reagent blank?

We want everything that might contribute to an absorbance at 595 nm *except* the protein we are trying to measure. Therefore, the best reagent blank is a tube of Bradford reagent without any added protein. By zeroing the machine on this tube, the absorbance due to the Bradford reagent is subtracted out automatically.

3.3 Standard Curves

Determining the concentration of a substance as in Practice Session 3.1 works well *if* you know the extinction coefficient and *if* you know that the system obeys Beer's law at that concentration. When these things are not known, which is most of the time, a standard curve is prepared. A standard curve is a plot of *A* versus a varying amount of a substance. Then, an unknown concentration can be determined from the graph.

PRACTICE SESSION 3.3

We have a compound X of varying concentrations in a phosphate buffer, pH 7.0, and the following absorbencies:

Tube No.	Concentration (mM)	Absorbance	Corrected Absorbance
1	0	0.05	0.00
2	1	0.15	0.10
3	2	0.25	0.20
4	3	0.35	0.30
5	4	0.45	0.40
6	5	0.55	0.50

What is the concentration of a sample of X if the absorbance equals 0.30?

First, you must understand **corrected absorbencies.** Tube 1 has an absorbance of 0.05, but it does not contain any of the compound that we are measuring. This tube is our reagent blank, and it has an absorbance. The graph must have a curve that goes through zero. There are two ways of dealing with this. The first way is to subtract the absorbance (0.05) from all the absorbencies. This gives the data shown in the last column. The second way is to zero the spectrophotometer with tube 1. That way, the subtraction is done automatically. We will always use corrected curves even though the difference between a corrected curve and an uncorrected one is largely cosmetic.

Continuing with the example, first subtract the reagent blank (tube 1) from the rest to give the corrected absorbance; then plot A versus c (corrected) (Figure 3.6). On the graph, look for the A that corresponds to the unknown, $0.30 - 0.05$ (blank) $= 0.25$. From the graph, $A = 0.25$ corresponds to the concentration of 2.5 mM, which is the answer. ●

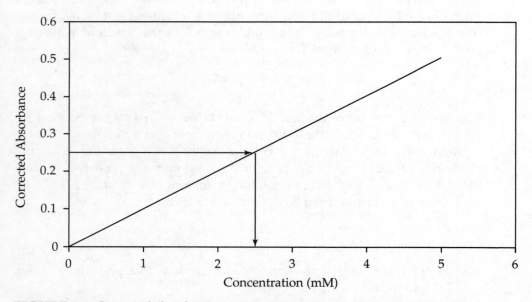

FIGURE 3.6 *Corrected absorbance versus concentration for Practice Session 3.3*

To make and use a standard curve, set up a series of tubes with varying amounts of substance in them. After measuring the absorbance, plot the corrected absorbance versus the amount put in. Sometimes you will plot absorbance versus concentration, but the most useful value to plot is either a weight (such as micrograms or milligrams) or an amount in a molar-based unit, such as micromole, millimole, and so on. Once you have your standard curve, use it to determine the concentration of an unknown. The absorbance value of the unknown tube must fall within the line of your standard curve, preferably within the linear region. Extrapolating the line beyond your highest concentration standard is not permitted.

3.4 Protein Assays

One of the most common uses for spectrophotometry, which also happens to use standard curves, is the protein assay. Many biochemical studies at some point require the knowledge of the amount of protein that you have in a sample.

Ultraviolet Absorption

Proteins can be assayed easily if you have a spectrophotometer that can measure light in the ultraviolet (UV) region. The amino acids tryptophan and tyrosine absorb strongly at 280 nm, which enables the scientist to scan for proteins at this wavelength. This is often done as a protein is purified with some chromatographic technique. However, to get a quantitative answer, you have to know the exact ε_{280} for the protein. If the protein contains few aromatic residues or the extinction coefficient is low, the UV method would not be suitable. Also, many spectrophotometers found in teaching labs do not have UV capability.

Colorizing Reagents

Many assays can compensate for the inability to use UV absorption. Most of them depend upon certain dye molecules that react with parts of the protein to give a colored complex that can then be measured. Once you have a colored complex, you can use visible light spectrophotometry.

One of the most common and easiest to use is the Bradford method. This method uses a dye called Coomassie Brilliant Blue G-250, which has a negative charge on it. The dye normally exists in a red form that absorbs light maximally at 465 nm. When the dye binds to the positive charges on a protein, it shifts to the blue form, which absorbs maximally at 595 nm. Many proteins have the same response curve to this dye, making the Bradford method reproducible among many experimenters. It is also very rapid. The reaction occurs in a couple of minutes, and the colored product is stable for over an hour. In addition, the protocol calls for a protein sample of up to 100 µL to be added to 3–5 mL of Bradford reagent. With such a large difference in volumes between the sample and the protein reagent,

TIP 3.1   Here are a few tricks to getting good results with a Spectronic 20:

1. Zero the machine often and close to the time when taking measurements.

2. Make sure that over 2 mL of solution are in the cuvette.

3. Be careful with the number of cuvettes that are used. The best data come from the fewest number of cuvettes.

bringing all samples up to the same 100-μL volume is not necessary before reagent addition. This saves time in setting up the assays.

3.5 Why Is This Important?

No single piece of equipment is used more in biochemistry or any life science than the spectrophotometer. We measure almost everything using one. Granted, most research labs have very sophisticated spectrophotometers that do more work, but basically it has a simple light source, a prism or grating for controlling the wavelength, and a sample holder. Getting caught up in the fun of pushing buttons without understanding how the machine works is far too easy. With any piece of equipment, the numbers generated only have meaning if the machine was calibrated and used properly. Most undergraduate labs have the Spectronic 20, Spectronic 20 D, Spectronic 20 D+ , or Spectronic 21, which you will use constantly. Take the time to really learn how to use them now. The payoff will be great.

3.6 Expanding the Topic

Calculating Concentrations from Graphs

A major source of frustration to many students is figuring out just what to plot against absorbance for these types of experiments. Two quantities can be plotted. The first is the final concentration of the product in the reaction vessel in molar, millimolar, milligrams per milliliter, percentage, and so on. The second is the amount of product put into the reaction vessel in milligrams, grams, millimoles, moles, and so on. The latter is more often the most useful because what we want to determine is the concentration of the unknown solution *before* it was put into the reaction vessel. As an example, consider the determination of creatinine. The objective is to determine the [creatinine] in the unknown serum. A hypothetical table might look like this:

	Reagent/Tube					
	1	2	3	4	5	6
Creatinine (1 mg/mL)	0	0.5	1.0	2.0	3.0	—
Water	3	2.5	2.0	1.0	0	2.0
Serum	—	—	—	—	—	1.0
Picrate	3	3	3	3	3	3
A_{500}	0	0.1	0.2	0.4	0.6	0.5

For simplicity, let's give the blank an absorbance A of zero (we should be so lucky). Now let's calculate the concentration of the unknown by the two methods. Keep track of the steps involved.

Method 1 (Not Recommended) We will plot A versus final [creatinine], so we need to calculate the concentrations. For each tube,

$$1\,\text{mg/mL creatinine} \times \text{mL added} = \text{mg creatinine added}$$

$$\frac{\text{added mg creatinine}}{6\,\text{mL}} = \text{final concentration}$$

	Tube					
	1	*2*	*3*	*4*	*5*	*6*
Creatinine (mg)	0	0.5	1	2	3	—
Final (mg/mL)	0	0.083	0.167	0.333	0.5	—

Then we plot A versus [creatinine], such as in Figure 3.7. From the graph, we find that $A_{500} = 0.5$ gives a concentration of 0.417 mg/mL. Each tube has 6 mL, so

$$0.417\,\text{mg/mL} \times 6\,\text{mL} = 2.5\,\text{mg creatinine}$$

FIGURE 3.7 *Absorbance versus [creatinine]*

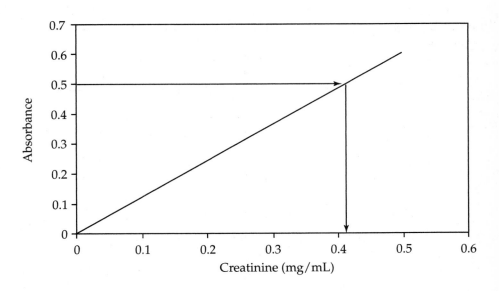

The 2.5 mg of creatinine came from 1 mL, so the concentration is

$$\frac{2.5\,\text{mg}}{1\,\text{mL}} = 2.5\ \text{mg/mL}$$

With this method, at one point you divided by 6 mL and then, a few steps later, multiplied by 6 mL.

Method Two (Recommended) This time we plot A_{500} versus mg creatinine added, using the data in the following table:

	Tube					
	1	*2*	*3*	*4*	*5*	*6*
Creatinine (mg)	0	0.5	1	2	3	—

Figure 3.8 shows the graph. From the graph, we find that $A_{500} = 0.5$ gives 2.5 mg of creatinine added. Because it came from 1 mL, we again calculate that 2.5 mg/mL of creatinine is in the serum.

This method eliminates the needless conversion to concentration within the reaction vessel and then back again. To help remember calculations like this, ask yourself, "What am I trying to figure out?" With assays designed to determine the concentration of something, the concentration in the assay tube is usually irrelevant. It is the concentration in the original sample that you want to know.

FIGURE 3.8 *Absorbance versus mg creatinine*

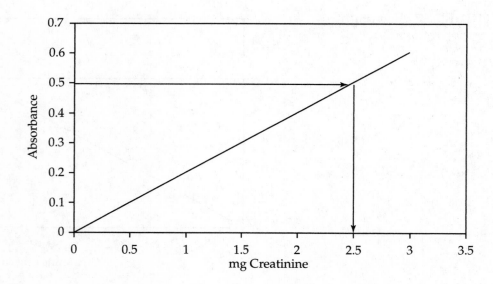

3.7 Tricks of the Trade

Choosing Test Tubes

How can you tell what test tubes to use? Anytime that you are going to be vortexing solutions in a tube, it should be half full or less. Otherwise, your vortexing will be ineffective at best and messy at worst.

When should you use just one cuvette? In a standard curve for a protein assay and other similar assays, setting up the reagents in large tubes is best and then pipetting or pouring some into one cuvette. This will accomplish two things. First, it prevents mixing reagents in tubes that are too small and too expensive to use for that purpose. Second, repeatedly reading the same tube is more accurate so that differences in the tubes do not confuse differences in the solutions. In other words, *you want as few variables as possible.*

Choosing Pipets

Pipets are at their best when used to capacity. You do not want to pipet 1 mL with a 10-mL pipet; you will have a huge error. There is a chance of error anytime that you pipet. The error grows as less and less of the pipet's capacity is used. It also grows as you use a pipet again and again; for instance, you do not want to pipet 5 mL by using a 0.2-mL pipet 25 times.

For the experiments in this chapter, you have a choice of instruments for pipetting the BSA into the tubes. The most accurate ones are the Hamilton syringe and the disposable micropipet. There are also micropipets of 5, 10, 20, 50, and 100 μL. The Hamilton is the most accurate pipet available. Its advantage is its accuracy, but the disadvantage is that it has to be cleaned. Also, if the needle or the plunger bends, it is permanently broken. The Pipetman can also be used. Its advantage is replaceable tips; its disadvantage is that it must be calibrated often and is not very accurate in untrained hands.

Experiment 3

Beer's Law and Standard Curves

In this experiment, you will learn how to use a spectrophotometer to calculate concentrations, using Beer's law. You will also do a simple protein assay and learn to create standard curves for the determination of the protein concentration in an unknown.

Prelab Questions

1. If your spectrophotometer can measure an absorbance up to 1.5, what is the maximum concentration of NADH that you can measure without diluting?

2. If you add 3 mL of water to 1 mL of NADH, mix and get an absorbance of 0.2, what is the concentration of the original NADH solution?

3. What size test tubes will you use to mix the reagents for the Bradford protein assay?

Objectives

Upon successful completion of this lab, you will be able to

- Zero the spectrophotometer at a variety of wavelengths and measure the absorbencies of solutions.

- Decide when dilutions must be made and make the appropriate dilutions.

- Calculate absorbencies, concentrations, and extinction coefficients, using Beer's law.

- Make standard curves and determine concentrations from them.

- Define a reagent blank and decide what reagents must be present in one.

- Design protocols for the creation of a standard curve.

Experimental Procedures

Materials

Spectrophotometers and cuvettes

NADH unknowns

Bradford protein reagent

BSA (bovine serum albumin), 1 mg/mL

BSA of unknown concentration

Methods

Part A: Using Beer's Law to Determine Concentration

In this part of the experiment, you use Beer's law to determine the concentration of a solution of NADH, a reagent used later in the course when doing enzyme purification. NADH has an extinction coefficient of $6220 \, M^{-1} \, cm^{-1}$ at 340 nm. The pathlength for the cuvette is 1 cm.

1. Obtain a solution of NADH of unknown concentration.

2. Warm up and zero the spectrophotometer at 340 nm or at the wavelength as close to 340 nm that you can. Sometimes older machines cannot be zeroed at 340 nm but can be zeroed somewhere between 340 and 360.

3. Make a minimal dilution of the NADH to provide enough solution to measure in the cuvette.

4. Measure the absorbance of the NADH. If the absorbance is greater than 0.8, dilute it with water and remeasure. Record these dilutions. You need to know how much NADH you added to how much water.

5. Use the absorbance and any dilutions you made to determine the concentration of the NADH millimolar (mM).

Part B: Setting Up a Standard Curve

In this part, you set up a standard curve for a protein determination, called the Bradford method. The protein standard is bovine serum albumin (BSA), a generic protein generally used for protein assays. Usually, a series of tubes is set up with varying amounts of BSA and a constant amount of Bradford reagent. By plotting the milligrams or micrograms of BSA on the x axis and the corrected absorbance on the y axis, we can determine the concentration of unknown proteins from the graph.

1. Warm up the spectrophotometer at 595 nm.

2. Set up ten large, clean test tubes to use for the assays. As a general rule, it is better to use large tubes for the reactions and then pour a couple milliliters of the solution into a cuvette-sized tube to read it, rather than setting up the reaction in the cuvettes. Cuvettes are too small to mix most reaction solutions, and you also risk permanently discoloring the cuvettes.

3. Set up a protocol as in Table 3.1. Using the most accurate pipet available, pipet the BSA standard into the tubes. The protein concentration is very high, and the volume is low, so any pipetting error will lead to poor standard curves.

4. Obtain an unknown BSA solution. Choose a volume of the unknown to assay and pipet into tube 9. This volume must be 100 μL or less.

5. Add 5 mL of Bradford reagent to each tube. Vortex immediately after adding the reagent to each tube. Do not wait until you have added it to all tubes.

6. Let the tubes sit about 10 min before reading the absorbencies. Once the color develops, it is stable for over an hour.

7. Make a plot to determine how much BSA you can add without the curve straying from linear. You may find that it was linear through your highest standard (100 μL), but it may also veer off at a lower volume of standard. If the absorbance of the unknown BSA is higher than your highest standard point that is on the straight part of the curve, you will need to make up another tube using less of the unknown BSA. You also want the corrected absorbance of your unknown to be 0.8 or less.

TABLE 3.1 *Protocol for Protein Determination*

	1	*2*	*3*	*4*	*5*	*6*	*7*	*8*	*9*
				Reagent/Tube (mL)					
BSA standard, 1 mg/mL (μL)	0	10	20	30	40	50	75	100	0
Unknown protein	0		0	0	0	0	0	0	0	?
Bradford reagent	3.0 mL				⟶				

Analysis of Results

Experiment 3: **Beer's Law and Standard Curves**

Data

Part A: Beer's Law

1. Unknown # _____

2. Dilution of NADH _____

3. Describe how you arrived at this dilution:

4. Absorbance of NADH _____

Part B: Bradford Protein Assay

1. Unknown # _____

2. Fill in the following table for your raw data.

Reagent/Tube	1	2	3	4	5	6	7	8	9	10	11
BSA volume (µL)	0	10	20	30	40	50	75	100	—	—	—
Unknown (µL)	—	—	—	—	—	—	—	—			
Protein (µg)											
Absorbance (corrected)											

Analysis of Results   *Part A: Beer's Law*

1. Concentration of NADH _____

2. Describe how you arrived at this number:

Part B: Bradford Protein Assay

1. Using graph paper, make a graph of corrected absorbance versus μg protein and attach it to this report.

2. Calculate the micrograms of protein in your unknown assays.

3. Calculate the concentration of the unknown protein by dividing the weight of protein you determined in step 2 by the volume of sample you put into the Bradford reagent.

unknown concentration = _____ μg ÷ _____ μL

= _____ mg/mL

Questions

1. What is the theoretical absorbance at 340 nm of a 0.01 M solution of NADH, assuming a 1-cm pathlength?

2. What dilution would be necessary to get the absorbance from Question 1 down to 3.1? (_____ mL of 0.1 M NADH to _____ mL H_2O)

3. Absorbance at 340 nm of a 0.02 mM NADH solution is 0.124 with a 1-cm pathlength. What is the absorbance with a 1.2-cm pathlength?

4. Five μL of a 10-to-1 dilution of a sample were added to 5 mL of Bradford reagent. The absorbance at 595 nm was 0.78 and, according to a standard curve, corresponds to 0.015 mg of protein on the x axis. What is the protein concentration of the original solution?

5. Why did we not use Beer's law in Part B?

6. How would your results have been affected if you neglected to change wavelengths between Part A and Part B?

7. Why is the absorbance versus concentration curve for a substance rarely straight for all concentrations?

Experiment 3a

Protein Concentration of LDH Fractions

In this experiment, you will learn to do the Bradford protein assay, a rapid and simple assay for protein concentration. This allows you to calculate the protein concentration of the lactate dehydrogenase (LDH) fractions from your purification experiments, which in turn allow you to calculate the specific activity and the fold purification at each step.

Prelab Questions

1. Why do you want to use large test tubes for the Bradford assay?

2. With the standard Bradford assay, why is it *not* necessary to equalize *all* sample volumes before adding the Bradford reagent?

3. Why is it not necessary to use a buffer to dilute your protein samples?

Objectives

Upon successful completion of this lab, you will be able to

- Make standard curves and determine protein concentrations from them.
- Design protocols for the creation of a standard curve.
- Complete the LDH purification table for Chapter 7.

Experimental Procedures

Materials

Spectrophotometers and cuvettes

Bradford protein reagent

BSA (bovine serum albumin), 1 mg/mL

LDH fractions from purification experiment

Methods

1. Turn on the spectrophotometer and set the wavelength to 595 nm.

2. Prepare *large* test tubes for your standard curve and your unknowns, the latter of which will be all of your saved LDH fractions.

3. Using the BSA standard, put enough BSA in the tubes to give a range from 0 to 100 μg of protein.

4. In some more tubes, put in a volume of your fractions. Because you do not know the concentration of these samples, just guess how much to put in, keeping in mind that the maximum volume is 100 μL for this assay. As a hint, the crude homogenate and anything that is visibly cloudy (that is, those fractions before the column chromatography) are very concentrated, so make a 10-to-1 dilution of those right away and try assaying 10 or 20 μL. It is always a good idea to assay duplicate tubes. Any duplicates that do not give the same number should be assayed again.

5. Add 5 mL of Bradford reagent to the tubes and vortex immediately. Let the tubes sit for 5 min.

6. While the tubes are sitting, zero the spectrophotometer at 595 nm, using tube 1 (zero protein) as the blank.

7. Assay the reactions by pouring roughly 3 mL of solution into a cuvette and reading the absorbance. Pour the solution back out into the large tube as soon as you have recorded the data.

8. Note the absorbance of your 100-μg standard. Any of your LDH samples that have an absorbance higher than that must be reassayed by making up a new tube with less fraction in it until it falls within your standard curve.

9. Calculate the protein concentration of your samples, the specific activity, and the fold purification, making the final purification table.

10. Before you leave, it is best if you go over a complete set of calculations for one fraction with the teaching staff.

Analysis of Results

Experiment 3a: **Protein Concentration of LDH Fractions**

Data

1. Fill in the following table for your standard curve:

Reagent/Tube	1	2	3	4	5	6	7	8
BSA volume (μL)	0	10	20	30	40	50	75	100
Protein (μg)								
Absorbance (corrected)								

2. Fill in the following table for your LDH fractions.

Fraction	Dilution Used	Volume Assayed	A_{595}	Protein (μg)	mg/mL
Crude					
20,000 × g, supernatant					
65% AS pellet					
Dialyzed 65% pellet					
Pooled IEX* fractions					
Dialyzed IEX fractions					

(continued)

Fraction	Dilution Used	Volume Assayed	A_{595}	Protein (μg)	mg/mL
Pooled Cibacron Blue fractions					
Concentrated Cibacron Blue fractions					
Pooled Sephadex fractions					
Concentrated Sephadex fractions					

* IEX is ion-exchange.

Analysis of Results

1. Using graph paper, make a graph of corrected absorbance versus μg protein and attach it to this report.

2. Does the trend in your protein concentrations make sense? Why or why not?

3. Complete the purification table in Chapter 7. The protein concentration (in mg/mL) is divided into the relative activity to give specific activity.

Additional Problem Set

1. The extinction coefficient for NADH is $6220\,M^{-1}\,cm^{-1}$ at 340 nm. Calculate the following:

    a. The absorbance of a $2.2 \times 10^{-5}\,M$ sample in a 1-cm cuvette at 340 nm

    b. The absorbance of a $2.2 \times 10^{-5}\,M$ sample in a 1-mm cuvette at 340 nm

    c. The absorbance of a 1 mM sample in a 1-cm cuvette at 340 nm

2. How would you calculate the extinction coefficient for NADH at 260 nm?

3. Define the terms *transmittance, percent transmittance, absorbance,* and *extinction coefficient.*

4. Calculate the molar extinction coefficient of a biomolecule with a molecular weight of 300 if a 5.0 mM solution in a 1.2-cm cuvette has an absorbance of 0.50 at 340 nm.

5. A compound has an extinction coefficient of $22{,}150\,M^{-1}\,cm^{-1}$. If a solution of this compound has an absorbance of 0.6 in a 1.15-cm cuvette, what is the concentration of the solution?

6. Aliquots of a 0.5-mg/mL standard of BSA are used to construct a standard curve for the Bradford protein assay. The tubes contain the following amounts of the BSA solution: 0, 20, 40, 60, 80, and 100 μL. The corresponding absorbencies after adding Bradford reagent are 0, 0.05, 0.09, 0.14, 0.19, and 0.22. If you take 20 μL of an unknown and add 80 μL of water, mix, take 10 μL of the mixture and add to Bradford reagent, and see an absorbance of 0.08, what is the protein concentration of the undiluted unknown?

Webconnections

For a list of web sites related to the material covered in this chapter, go to **Webconnections** at the *Experiments in Biochemistry* site on the Brooks/Cole Publishing web site. You can access this page at http://www.brookscole .com and follow the links from the chemistry page.

References and Further Reading

Bradford, M. M. *Analytical Biochemistry* 72 (1976): 248.

Burgess, C., and K. D. Mielenz. *Advances in Standards and Methodology in Spectrophotometry.* Elsevier Science, 1987.

Gorog, S. *Ultraviolet and Visible Spectrophotometry in Pharmaceutical Analysis.* CRC Press, 1995.

Jack, R. C. *Basic Biochemical Laboratory Procedures and Computing.* New York: Oxford University Press, 1995.

Plummer, D. T. *An Introduction to Practical Biochemistry*, 2nd ed. London: McGraw-Hill, 1978.

Robyt, J. F., and B. J. White. *Biochemical Techniques*. Long Grove, IL: Waveland Press, 1990.

Segel, I. H. *Biochemical Calculations*, 2nd ed. New York: Wiley Interscience, 1976.

Stenesh, J. *Experimental Biochemistry*. Boston: Allyn and Bacon, 1984.

Sommer, L. *Analytical Absorption Spectrophotometry in the Visible and Ultraviolet: The Principles*. Elsevier Science, 1990.

Chapter 4

Enzyme Purification

Introduction

Enzymes control all biochemical reactions in an organism. To understand the metabolism of a cell or organism, one must understand the enzymes responsible. Before studying an enzyme, it must be purified away from contaminants. In this chapter, we begin the study of enzyme purification. The enzyme lactate dehydrogenase will be purified from beef heart tissue via several procedures. We begin with the procedures of homogenization, centrifugation, and ammonium sulfate precipitation.

4.1 Enzymes as Catalysts

Although most of the reactions that occur in the human body are possible in thermodynamic terms, they take place far too slowly to be useful without a catalyst. Enzymes are the biological catalysts that allow metabolism to happen. Without them, the only way for the same reactions to occur is to raise the temperature to extremely high levels in which life could not exist. Enzymes work by lowering the energy of the transition state between the reactants and the products. Figure 4.1 shows the energy diagram of this effect.

With the exception of some recently discovered RNAs, all enzymes are proteins. These proteins are the most efficient catalysts known. Enzymes can increase the speed of a reaction to 10^{20} times that of an uncatalyzed reaction. These proteins are also very specific for the substrates of the reaction. If an enzyme normally reacts with L-alanine, for example, it may not recognize D-alanine at all. Some other enzymes, however, may have several substrates that they recognize and react with.

4.2 Enzyme Purification

Biological systems are very complex and difficult to study *in vivo*, so enzymes are usually separated from other proteins and metabolites and studied individually. This separation is called **enzyme purification**, and

FIGURE 4.1 *Activation profiles for catalyzed and uncatalyzed reactions*

(a)

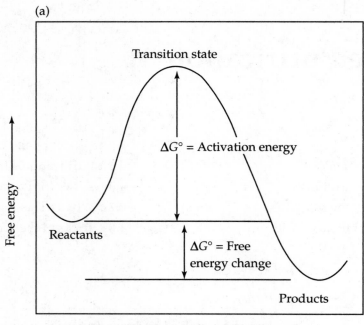

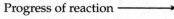

(b)

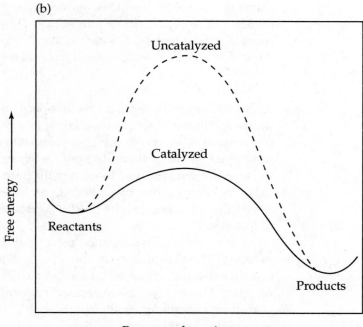

the desired product is a preparation that contains only the enzyme of interest. Usually, many techniques are combined, with each one removing more of the unwanted material.

A common way to purify an enzyme is to take a tissue source rich in enzyme and grind it in a buffer. This is called **homogenization.** From that moment on, you are in a race against time. Inside the cell, the enzyme of interest may be protected from other enzymes whose job it is to degrade proteins. Once the tissue is homogenized, these degradative enzymes, called **proteases,** begin to degrade all proteins. Biological enzymes function best at around body temperature, or 37°C. If the solution is kept cold—say, around 4°C—then we can purify the enzyme and keep the proteases from functioning. That is why most enzyme purifications are done in a cold room or on ice the entire time. You hope that as you purify, you do not lose too much of the enzyme but do lose most of the contaminating proteins from the homogenate.

In the experiments that follow, we will partially purify lactate dehydrogenase (LDH), which we will use for our model purification.

Homogenization of LDH

We will use beef heart as our source of LDH—an enzyme found in the cell cytosol, which makes it easy to isolate. The procedure is as follows: Cut some heart muscle into small pieces and trim off any obvious nonmuscle tissue, such as fat. Then mix the heart pieces with a suitable buffer, usually a phosphate buffer at pH 7.5, in a prescribed ratio. Next a blender grinds the heart tissue in a 4°C cold room. Now we have a beef heart homogenate suitable for purifying.

Centrifugation

Sometimes particles are separated from others based on differences in density. When particles are spun in a centrifuge, they will move according to their density, the density of the solution, and the force applied. For example, in a crude tissue homogenate, if the force applied is $500 \times g$, unbroken cells and nuclei will precipitate while the other particles will stay in solution. If the force is increased to $10,000 \times g$, mitochondria, peroxisomes, and lysozomes will spin down. To precipitate the microsomes (fragments of endoplasmic reticulum), use $100,000 \times g$. Figure 4.2 shows the fundamentals of separating subcellular organelles by centrifugation. Other times, the centrifuge is used to precipitate everything to the bottom.

TIP 4.1 When purifying an enzyme, always remember that every second that your sample is not in a cold room or an ice bucket it is susceptible to degradation. The difference between your enzyme preparation and the student next to you is usually the care it is given in-between the actual purification steps.

FIGURE 4.2 *Differential centrifugation*

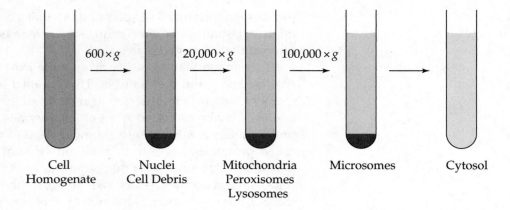

To control the force applied, it is necessary to know the radius of the centrifuge rotor. The larger the radius, the greater is the force at a given speed. Usually, centrifugations are reported in terms of forces × *g*. You must then calculate how fast to run it to get the desired force. Tables exist to convert RPMs (revolutions per minute) to *g* forces for different rotors. Table 4.1 shows a typical conversion chart for rotor speed to force × *g*.

In the purification of LDH, centrifugation will be used three times. The first time is when the crude homogenate is spun at 20,000 × *g*. This

TABLE 4.1 *Rotor Speed versus g Force*

RPM	JA-21 102	JA-20 108	RPM	JA-21 102	JA-20 108
500	28	30	8,000	7,310	7,740
1,000	114	120	8,500	8,250	8,740
1,500	257	272	9,000	9,250	9,800
2,000	456	483	9,500	10,300	10,900
2,500	714	756	10,000	11,400	12,100
3,000	1,030	1,090	10,500	12,600	13,300
3,500	1,400	1,480	11,000	13,800	14,600
4,000	1,830	1,940	11,500	15,100	16,000
4,500	2,310	2,450	12,000	16,500	17,400
5,000	2,860	3,030	12,500	17,900	18,900
5,500	3,460	3,660	13,000	19,300	20,400
6,000	4,110	4,350	13,500	20,800	22,000
6,500	4,830	5,110	14,000	22,400	23,700
7,000	5,600	5,930	14,500	24,000	25,400
7,500	6,430	6,800	15,000	25,700	27,200

precipitates out any unbroken cells, nuclei, mitochondria, and most of the peroxisomes and lysozomes. No LDH should be in any of these except the unbroken cells because LDH is a cytosolic enzyme. Once you have collected the supernatant from the $20,000 \times g$ spin, add ammonium sulfate and a precipitate forms. Centrifugation spins down this precipitate, which should have only contaminating proteins. To the supernatant from that spin, add more ammonium sulfate. This forms another precipitate, which should have the LDH. This then is spun down and collected.

Salting Out

From the $20,000 \times g$ supernatant solution, which contains many contaminating proteins, we must purify the enzyme. A common way is to start with **ammonium sulfate precipitation.** This technique is called **salting out.** Proteins are kept in solution by interactions between their hydrophilic portions and the solvent. Hydrogen bonds form and the proteins are surrounded by the water molecules in the solvent. By adding a very polar compound—such as ammonium sulfate, $(NH_4)_2SO_4$—many of the water molecules will interact with the salt instead of the proteins. With less water available to stabilize the proteins, the proteins begin to interact with each other and clump together, much like oil does when mixed with water.

Certain proteins precipitate out of solution at low concentrations of ammonium sulfate, whereas others require higher concentrations. We will add a low concentration of ammonium sulfate to the homogenized beef heart, allow the proteins to precipitate, and centrifuge to drive the precipitated proteins to the bottom. To the supernatant, we will add more ammonium sulfate to precipitate the LDH. This will then be centrifuged, and the pellet collected.

When doing an ammonium sulfate precipitation, it is best to use either a saturated solution or ammonium sulfate crystals that have first been ground into a powder. This allows for quicker mixing without getting locally high concentrations of the salt, which will cause the wrong proteins to precipitate out. The powder must be added slowly with good mixing for the same reasons.

Further Purification Steps

If you want to purify LDH to 100% purity, you have to choose several other techniques. A good combination uses several chromatography techniques, such as ion exchange (Chapter 5), affinity chromatography (Chapter 6), or gel filtration (Chapter 7). Each of these column techniques

TIP 4.2 The best way to get a good ammonium sulfate precipitation is to add the salt after it has been ground into a powder because it goes into solution more easily. Also add it very slowly so that you never have a locally higher concentration.

provide a sample of increasing purity. In Experiments 4 and 4a, however, we stop at the ammonium sulfate precipitation.

4.3 Units of Enzyme Activity

Before beginning to purify, you must understand the quantities that will be measured. Enzymes catalyze a specific reaction, and what we are interested in is the amount of product formed per unit time. This is called the **activity:**

$$\text{activity} = \frac{\textbf{amount of product}}{\textbf{time}}$$

Usually, enzyme activity is measured in micromoles of product per minute. This would be 1 **unit of activity:**

$$\textbf{1 unit of activity (U)} = \textbf{1 } \mu\textbf{mol/min}$$

However, we can define a unit to be anything we want, depending on the assay used. If you are studying an enzyme with extremely low activity in the cell, you might tire of writing 0.003 unit, so you can define 1 unit to be 1 nmol/min and will have 3 units instead. Thus, when measuring the activity of an enzyme, we are not actually measuring the number of enzyme molecules but are measuring what those molecules are *doing*. The activity is very dependent on reaction conditions—such as pH, ionic strength, temperature—and the presence of inhibitors or activators.

If we want to know the concentration of an enzyme sample, then we calculate how many units of activity are in a certain volume. This is called **relative activity** and is usually measured in units per milliliter:

$$\textbf{relative activity} = \textbf{units/mL} = \textbf{U/mL}$$

Therefore, if we use the preceding definition of units and add 0.10 mL of an enzyme preparation and find that the reaction proceeds at 10 μmol/min, the relative activity of the enzyme preparation is

$$\frac{10 \ \mu\text{mol/min}}{0.1 \ \text{mL}} = \frac{100 \ \mu\text{mol/min}}{\text{mL}} = 100 \ \text{U/mL}$$

The number of units per milliliter tells us the concentration of the enzyme, but it does not tell us its purity. What we want is a high activity of the enzyme without a lot of other proteins around. To measure this, we use **specific activity,**

$$\textbf{specific activity} = \textbf{U/mg protein}$$

which is the activity in units divided by the milligrams of total protein in the sample. Keep in mind that you can have two fractions of equal specific activity that are quite different. Because specific activity is a ratio, anything that increases the number of units or decreases the number of milligrams of protein increases the specific activity. The number of milligrams of protein is a measure of total protein, active or inactive, enzyme or nonenzyme. The number of units is only the live form of the enzyme we want. In general, the fraction with the highest specific activity is considered the most pure.

PRACTICE SESSION 4.1

We have three enzyme fractions. We take $200\,\mu$L of each and assay for activity. We also take $0.5\,$mL of each and assay for protein. The results are the following:

Fraction	Units	Protein (mg)
1	10	10
2	20	10
3	25	20

Calculate the specific activity of each fraction.

For fraction 1, $10\,$U came from $200\,\mu$L, or $0.2\,$mL; therefore, $10\,$U$/0.2\,$mL $= 50\,$U$/$mL. Furthermore, $0.5\,$mL of the fraction contains $10\,$mg of protein, so $10\,$mg$/0.5\,$mL $= 20\,$mg$/$mL.

$$\text{specific activity} = \text{U/mg} = \frac{\text{U/mL}}{\text{mg/mL}} = \frac{50}{20} = 2.5\ \text{U/mg}$$

Following the same logic for the other fractions gives $5\,$U$/$mg and $3.125\,$U$/$mg, respectively.
●

Two other important quantities are the **percent (%) recovery** and the **fold purification.** There is always a starting point in a purification; as you purify, compare the fractions to the starting point. The starting point is usually called the crude.

$$\textbf{\% recovery} = \frac{\text{total units of fraction}}{\text{total units of crude}} \times 100$$

That is, if we start with $1000\,$U in the crude and we have $500\,$U in the fraction, there is a 50% recovery at that point.

$$\textbf{fold purificaion} = \frac{\text{specific activity of fraction}}{\text{specific activity of crude}}$$

Therefore, if the purified fraction has a specific activity of 2000 U/mg and the crude has one of 100 U/mg, the fold purification is 20.

Usually, the fold purification increases while the percent recovery decreases during a purification. Sometimes, though, you may see strange things, like a percent recovery that is over 100%. Don't panic. This happens for two common reasons. First, you are comparing all your fractions to a crude sample. Crude samples are difficult to measure because they contain too many contaminating proteins and particulates, so your measurement of the crude may be the least accurate measurement you have. Second, remember that you are measuring an enzyme activity, *not* the number of enzyme molecules. Clearly, you cannot gain molecules of LDH, but you can gain LDH activity if each molecule suddenly became more active. This happens during a purification when an inhibitor is removed or the enzyme is put into a more favorable buffer or salt solution.

4.4 Calculating Initial Velocity

Whenever we assay an enzyme, we want to measure the initial velocity. This is the velocity or units that is calculated at the beginning of the reaction. The rate of reaction is dependent on the concentration of substrates. When it is the enzyme we want to measure, we try to hold the other variables constant by using large quantities of substrate. When the enzyme encounters the substrate, it converts it to products. The reaction eventually slows down because of substrate depletion and an opposing back reaction from the built-up products. The easiest way to measure the initial velocity is to graph the absorbance versus time for the reaction. If we measure the velocity of an enzyme-catalyzed reaction, we might see the data shown in Table 4.2.

If you graph these data, they appear as shown in Figure 4.3. Notice that the reaction rate is linear for the first minute but then decreases. Many students make the mistake of calculating their enzyme rates by taking the absorbance at 2 minutes, subtracting the absorbance at time zero, and then dividing by 2. This gives an absorbance change per minute, but it is *not* the initial velocity because absorbencies were included that were not on the linear portion. Doing the calculation that way indicates an absorbance change of $0.32/2 = 0.16$/min. This number is low; the true initial velocity is 0.2/min. Calling 0.16 your initial velocity is akin to admitting to the highway patrolman that your were going 85 mph on one stretch of highway but that he shouldn't ticket you because you were planning on only going 40 mph on the next stretch.

The safest way is to graph the data and take the initial slope of the line. This always works to give you the initial velocity. Another way is to look at the data in the table. Notice that the absorbance is changing 0.05 every 15 seconds and that this change is constant. From those data, you can conclude that your absorbance change is 0.2/min. Once you have calculated the initial velocity, you may convert to the other units mentioned in Section 4.3, such as relative activity, total activity, percent recovery, and so on. To do this, keep careful records regarding the amount of sample put into the cuvette to assay and any dilutions made previous to that. Without that information, you cannot do the calculations.

TABLE 4.2 *Absorbance versus Time*

Time (s)	Absorbance
0	0
15	0.05
30	0.10
45	0.15
60	0.20
75	0.24
90	0.27
105	0.30
120	0.32

FIGURE 4.3 *Absorbance versus time for enzyme-catalyzed reactions*

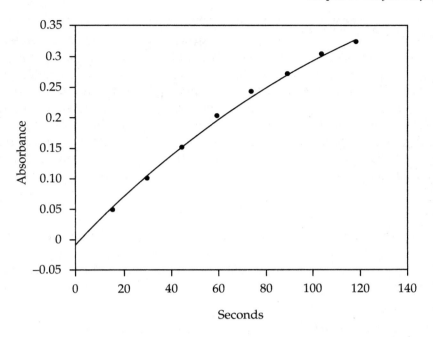

PRACTICE SESSION 4.2

Time (s)	Absorbance
0	0
15	0.05
30	0.10
45	0.15
60	0.20
75	0.24
90	0.27
105	0.30
120	0.32

If you are doing an enzyme assay that generates the data shown in Table 4.2, what is a good way to record your data?

The data in Table 4.2 are presented in a good way, but the information is incomplete. This might have been an assay of a particular fraction of your purification. A good way to present the data is the following:

Fraction: 40% ammonium sulfate supernatant

Sample volume: 20 μL

Dilution factor: 10/1

Total volume of fraction: 38 mL

Absorbance: 0.2/min

Initial velocity: 0.096 μmol/min

TIP 4.3 When doing enzyme purification experiments, always keep track of the initial volume of each sample, the volume of the sample put into the assay cuvette, and the dilution made before putting the sample in the cuvette.

ESSENTIAL INFORMATION

When studying enzymes, we must calculate how much enzyme we have. We do that by calculating the units of enzyme activity. This is an indirect method based on the measurement of the amount of product that the enzyme can produce from substrate per unit of time. We are in effect measuring what the enzyme does rather than what it is. To make this measurement, cal- culate a change in absorbance versus time for the reaction where substrate goes to product. Make sure to measure the rate at the beginning of the reaction, which is called the initial veloc- ity. The most secure way to do this is to graph absorbance versus time for each enzyme reac- tion and take the initial slope as your velocity. Then convert this to enzyme units.

4.5 Purification Tables

During the course of a purification, we usually make a purification table to help keep track of the success of the process. A purification table tracks all the quantities listed in Section 4.3 as we proceed from the crude homogenate to the final product. Table 4.3 shows a sample purification table.

4.6 Assay for Lactate Dehydrogenase (LDH)

LDH catalyzes the following reaction:

$$\text{pyruvic acid} + \text{NADH} \rightleftharpoons \text{NAD}^+ + \text{L-lactic acid}$$

We will measure the rate of reaction by monitoring the appearance of NADH, which absorbs light strongly at 340 nm. By watching the absorbance rising and timing how fast it is increasing, we can calculate how much NADH is being formed per time, which we can convert to units of enzyme activity.

We will be setting up a reaction cocktail containing an assay buffer, a lactate, and NAD^+. This reaction cocktail is clear, so you will have no visible way to know if it is correct or not. The starting absorbance of the cocktail should be around zero when compared to water. If it is much higher than that, the cocktail is faulty, probably due to NADH already produced.

TABLE 4.3 *Sample Purification Table*

Fraction	U/mL	Total Units	% Recovery	Protein (mg/mL)	U/mg	Fold Purification
Crude homogenate	45	2000	100	8.5	5.3	1
20,000 × g supernatant	45.5	1989	99	8.0	5.7	1.1
40% Ammonium sulfate supernatant	48	1824	91	7.0	6.9	1.3
65% Ammonium sulfate pellet	212	1224	61	12.6	16.8	3.2
Ion-exchange column- pooled fractions	186	1300	65	0.8	233	44

You must be *very careful* that any pipet that touched LDH never comes in contact with your reaction cocktail. Use a clean, new pipet tip every time.

Converting from Absorbance to Units

For the LDH assay, we use the extinction coefficient of NADH to convert from absorbance change per minute to units. Our standard assay has a volume of 3 mL (0.003 L). To report units in micromoles of product per minute, use the following equation:

$$U = \frac{(\Delta \text{ absorbance}/\Delta \text{ min})}{6220 \text{ M}^{-1} \text{ cm}^{-1} \text{ (cm)}} \times 10^6 \, \mu\text{M/M} \times 0.003 \text{ L} = \mu\text{mol/min}$$

The standard assay will not change during these experiments, so we can combine most of this information into a constant, which turns out to be 0.48:

$$U = \frac{\mu\text{mol}}{\text{min}} = \frac{\Delta \text{ absorbance}}{\Delta \text{ min}} \times 0.48$$

4.7 Why Is This Important?

We are using a well-known enzyme, LDH, to study centrifugation and enzyme purification. The most important thing you will get out of this is the ability to use a preparative centrifuge. Although this is not difficult to use, using it correctly is important. Most of what we know about metabolism came from the combination of enzyme purifications done on the individual enzymes responsible. You cannot study glycolysis or the Krebs cycle as a whole. The individual pieces must be studied.

The next time you are taking part in your favorite sport and you "feel the burn," you will know that lactic acid is being produced. I am sure that you will quickly say to yourself, "My legs are burning because I am utilizing glucose anaerobically and lactic acid is the final product, which makes my legs hurt," instead of just saying, "Ouch." Try it; it works.

You are also learning the early steps and theories behind classical enzyme purification. There is no magic to purifying an enzyme but is the combination of a logical sequence of steps designed to rid a sample of all contaminants. Starting with the simple techniques in this chapter and adding a few more from subsequent chapters, you will have a reasonable chance at purifying an enzyme to homogeneity.

4.8 Tricks of the Trade

Hints for Getting Good Enzyme Rates

1. Do not vortex enzymes. Invert the cuvette quickly but gently three times after adding the enzyme.
2. Have the cuvette with reaction cocktail warmed to room temperature before adding the enzyme.

3. Avoid using 100 μL of your samples because they are very crude and there will be much light scattering. Start with a 20-μL sample plus 80 μL of extra water and see if the rate is measurable.

Hints for Doing the Calculations Later

Whenever you do enzyme purifications, you must do many calculations. For each assay, some essential pieces of information are needed to do those calculations:

1. Record the volume of enzyme that went into the assay vessel. Note that this is not 100 μL each time. If 80 μL of water is added to the assay vessel and then 20 μL of the enzyme sample is added, the number recorded is 20 μL.

2. Record any dilutions you made. This is probably the most confusing part to those who do not do this for a living. Following the preceding hint, a dilution is not made just because 80 μL of water was added first and then 20 μL of enzyme. This is *not* a 5-to-1 dilution. We actually assayed 20 μL of the undiluted sample. A dilution is when a new, separate sample is made before putting some of it into the assay vessel.

   If we take 100 μL of 20,000 × *g* supernatant and add 900 μL of homogenization buffer, we have made a 10-to-1 dilution. If we then take 20 μL of that dilution and put it into an assay vessel with 80 μL of water and 2.9 mL of reaction cocktail, then we have assayed 20 μL of a 10-to-1 dilution.

3. Record the total milliliters of each fraction. You eventually will need to know that you had, for example, 35 mL of 20,000 × *g* supernatant, 40 mL of 65% supernatant, and so on.

Experiment 4

Purifying LDH (Short Version)

In this experiment, you will partially purify LDH from beef heart, using the techniques of homogenization, centrifugation, and ammonium sulfate precipitation. You will also learn how to do the assay for LDH.

Prelab Questions (Week 1)

1. What is the basis for the LDH assay?

2. How do you know if a reaction cocktail has been compromised?

3. What must be done just before putting the centrifuge tubes into the centrifuge?

4. Why do you add *ground* ammonium sulfate, and why do you add it *slowly*?

Prelab Questions (Week 2)

1. Draw the protocol you will use for the protein standard curve. How many milliliters of 0.5 mg/mL BSA will you use in each tube to get the range of 0–50 μg that you want?

2. What size and number of tubes will you use for the various parts of the experiment?

Objectives

Upon successful completion of this lab, you will be able to

- Explain simple purification schemes for enzymes.
- Explain how salting out with ammonium sulfate allows biological molecules to be separated.
- Utilize centrifugation for separation of biological preparations.
- Dilute enzyme preparations and assay for LDH activity.
- Construct standard curves and determine protein concentrations of enzyme fractions by the Bradford method.
- Determine the specific activity of enzyme fractions and construct a purification table for the purification of LDH.

Experimental Procedures

Materials

Beef heart crude homogenate

Assay buffer, 0.15 M CAPS, pH 10.0

NAD^+, 6 mM

Lactic acid, 150 mM

Ammonium sulfate

BSA (bovine serum albumin) standard solution, 0.5 mg/mL

Bradford reagent for protein determination

Methods

Lab Period 1

Carry out all procedures on ice.

1. Acquire 35 mL of ground beef heart in buffer and put it in a beaker on ice as quickly as possible. *Save 1 mL for assays and protein determinations before going on to step 2.*

2. Put the homogenate into centrifuge tubes or bottles as directed in lecture and balance with another student's bottle or with a bottle of water. *The balancing is of the utmost importance!*

3. Centrifuge at $20,000 \times g$ for 20 min. The centrifuges will be set at 4°C.

4. While the centrifugation is going, assay the crude homogenate for enzyme activity. The reaction cocktail for the assay will be the following:

   Assay buffer: 1.9 mL

   Lactate: 0.5 mL

   NAD^+ : 0.5 mL

This is the amount for one assay, so make up multiples of that depending on the number of assays you plan to do. Zero the spectrophotometer with water at 340 nm or as close to 340 as your machine will allow. Measure the absorbance of the reaction cocktail to be sure that it is close to zero.

5. To the 2.9 mL of cocktail, add extra assay buffer and then LDH from your crude homogenate to bring the volume up to 3 mL. Start with 10 μL of the sample. Invert three times and measure the absorbance changes immediately.

6. Use whatever dilutions are necessary to get an accurate change per minute. Dilute with assay buffer (*not* reaction cocktail or water). Be sure to record your dilutions.

7. Freeze the leftover crude homogenate sample for next week and save your dilutions.

8. When the centrifugation is over, discard the pellet and save the supernatant.

9. Measure the volume of the supernatant. *Save 1 mL for enzyme assay* and *slowly* add 0.230 g of ground (powder, not crystal) ammonium sulfate for every milliliter of supernatant. Stir the solution constantly while you add the salt. This amount of ammonium sulfate will bring the percent saturation to 40. This is called taking a 40% cut.

10. Let the solution stand (on ice as always) for 10 min while the proteins precipitate.

11. Balance the tubes and centrifuge at 15,000 × *g* for 15 min.

12. Discard the pellet. Save and measure the supernatant. Take 1 mL of the supernatant and assay for LDH activity as before.

13. To the rest of the supernatant, add 0.166 g of ground ammonium sulfate for every milliliter of supernatant. Add the salt slowly with constant stirring as before. This amount of ammonium sulfate will bring the final concentration up to 65% saturated.

14. Let the solution stand as before. While this is going on, assay the 40% supernatant that you saved.

15. Spin the 65% solution as before. Save the supernatant and pellet.

16. Resuspend the pellet in a small quantity of homogenization buffer (2–5 mL).

17. Measure the volume of the resuspended 65% pellet and assay for LDH activity. Assay the 65% supernatant.

18. Save all your samples in microfuge tubes (crude homogenate, 20,000 × *g* supernatant, 40% supernatant, 65% pellet, and 65% supernatant). These will be used next week for protein determinations.

19. At this point in the experiment, calculate the activity (units in the reaction cuvette), relative activity (units per milliliter of sample), and total activity (total units of sample) for each sample.

20. If this experiment is going to continue with Experiment 6, the resuspended 65% pellet will be put into a dialysis tube and dialyzed in 0.02 M sodium phosphate, pH 7.0.

Lab Period 2 (if Not Going on to Experiment 6)

1. Make a protein standard curve for the Bradford assay (see Chapter 3) using the 0.5 mg/mL BSA solution. You want 0–50 μg of BSA in the tubes. At the same time, set up protein assays for your five fractions from Lab Period 1. You will probably need small quantities, such as 10 μL, but you will have to play around with that.

2. Add 5 mL of Bradford reagent, vortex immediately, and let stand for 10 minutes. Read the absorbencies at 595 nm.

3. If your sample absorbencies are off of your standard curve, remake the tubes using less sample or a diluted sample.

4. At this point, calculate the specific activity (number of units/milligrams of protein) of the fractions, percent recovery (total units of fraction/total units of crude homogenate × 100), and the fold purification (specific activity fraction/specific activity crude homogenate).

Name _____   Section _____

Lab partner(s) _____   Date _____

Analysis of Results

Experiment 4: **Purifying LDH (Short Version)**

Data

Lab Period 1

1. Provide all information on your purification steps:

    Milliliters of crude homogenate used _____

    Milliliters of 20,000 $\times$ g supernatant _____

    Grams of $(NH_4)_2SO_4$ used for 40% cut _____

    Milliliters of 40% supernatant _____

    Grams of $(NH_4)_2SO_4$ used for 65% cut _____

    Milliliters of 65% supernatant _____

    Milliliters of 65% pellet after resuspending _____

2. Provide all information on your enzyme assays:

Fraction Isolated	Sample Assayed (µL)	Dilution Made (if any)	Absorbance Change per Minute
Crude homogenate			
20,000 $\times$ g supernatant			
40% $(NH_4)_2SO_4$ supernatant			
65% $(NH_4)_2SO_4$ pellet			
65% $(NH_4)_2SO_4$ supernatant			

Lab Period 2

1. Provide all information on your samples and the Bradford assay.

Standard Curve

Volume BSA	Absorbance

Bradford Assay on LDH Samples

LDH Fraction	Quantity Assayed (μL)	Dilution Used	Absorbance
Crude homogenate			
20,000 $\times$ g supernatant			
40% $(NH_4)_2SO_4$ supernatant			
65% $(NH_4)_2SO_4$ pellet			
65% $(NH_4)_2SO_4$ supernatant			

Calculations

1. For each of your fractions, calculate the activity in the reaction vessel (units in μmol/min) and the relative activity of the sample (units/mL of fraction). The extinction coefficient is $6220\,M^{-1}\,cm^{-1}$. The cuvettes are 1 cm in diameter.

$$\text{units} = \frac{\Delta A/\Delta \min}{6220\,M^{-1}\,cm^{-1}\,(1\,cm)} \times 10^6\,\mu M/M \times 3 \times 10^{-3}\,L$$

$$U/mL = \frac{\text{units from above}}{\text{volume of fraction assayed}} \times \text{dilution used, if any.}$$

2. Calculate the total units for each fraction by multiplying the relative activity by the total volume of the fraction.

3. Calculate the percent recovery of the purified fractions compared to the crude homogenate.

4. Calculate the protein concentration of the fractions in milligrams/milliliter using the Bradford assay standard curve. Attach your graph of absorbance at 595 nm versus μg protein.

5. Calculate the specific activity of each fraction by dividing the relative activity by the quantity of protein in milligrams/milliliter.

6. Calculate the fold purification by dividing the specific activities of the purified fractions by that of the crude homogenate.

All these calculations may be summarized in the following table.

Fraction	Units	Units/mL	Total Units	% Recovery	Protein (mg/mL)	Specific Activity	Fold Purification
Crude homogenate				100			1
20,000 × g supernatant							
40% $(NH_4)_2SO_4$ supernatant							
65% $(NH_4)_2SO_4$ pellet							
65% $(NH_4)_2SO_4$ supernatant							

Questions

1. Assayed for LDH activity were 5 μL of a sample that was diluted 6 to 1. The activity in the reaction vessel, which has a volume of 3 mL, is 0.30 U. What is the Δ Absorbance/minute observed? What is the relative activity of the original sample?

2. Of the original "undiluted" sample from Question 1, 10 µL of a 5-to-1 dilution were used to measure protein concentration. With the use of a standard curve, this is found to be 140 µg. What is the specific activity of the original sample?

3. If all students start with the same 10,000 $\times$ g supernatant, by the time different groups have done the 40–65% ammonium sulfate cut, their results usually vary greatly. What are likely reasons for this variance?

4. What is the difference between a percent saturated solution of ammonium sulfate and a percent w/v solution of ammonium sulfate?

5. What can you conclude about the effectiveness of these techniques for purifying LDH? How do your data support your conclusions?

Experiment 4a

Purifying LDH (Comprehensive Version)

In this experiment, you will begin a comprehensive experiment about protein purification. LDH will be isolated from beef heart by homogenization and centrifugation. It will be further purified by ammonium sulfate precipitation and dialysis.

Prelab Questions (Week 1)

1. What is the basis for the LDH assay?

2. How do we know if a reaction cocktail has been compromised?

3. What must be done just before putting the centrifuge tubes into the centrifuge?

4. Why do you add *ground* ammonium sulfate, and why do you add it *slowly*?

Prelab Questions (Week 2)

1. Draw the protocol you will use for the protein standard curve. How many milliliters of 0.5 mg/mL BSA will you use in each tube to get the range of 0–50 µg that you want?

2. Why do you need to mix the protein and Bradford solution immediately?

3. What can be an explanation if a tube containing 5 mL of Bradford reagent added to 10 μL of BSA standard showed a lower absorbance than the Bradford reagent blank?

Objectives

Upon successful completion of this lab, you will be able to

- Explain simple purification schemes for enzymes.
- Explain how salting out with ammonium sulfate allows biological molecules to be separated.
- Use centrifugation for separation of biological preparations.
- Dilute enzyme preparations and assay for LDH activity.
- Begin a ....fication table for the purification of LDH.

Experimental Procedures

Materials

Beef heart

Assay buffer, 0.15 M CAPS, pH 10.0

NAD^+, 6 mM

Lactic acid, 150 mM

Ammonium sulfate

Homogenization buffer, 0.05 M sodium phosphate, pH 7.0

Q-Sepharose buffer, 0.03 M bicine, pH 8.5

Methods

All of the following should be done in a cold room or on ice unless it just isn't possible. Start each experiment with about 100 mL of crude homogenate. Some steps may be done in groups to account for the number of rotor spaces and the size of centrifuge bottles being used.

At each step of a purification, it is important to record the volume of the sample that you are working on. Also, save 0.5 mL for enzyme assays and future protein determinations. After the enzyme assays, freeze the leftover samples for Experiment 3a.

Preparing LDH from Crude Tissue

1. Cut up the beef heart into fine pieces, avoiding the obvious fatty deposits and connective tissue.

2. Combine 25 g of beef heart with 75 mL of cold 0.05 M sodium phosphate, pH 7.0, and homogenize at high speed in a blender for 2 min at 4°C. *Save 0.5 mL for enzyme and protein assays.*

3. Spin the homogenate at 20,000 × g for 15 min at 4°C. *Save 0.5 mL of the supernatant for assays.* This will be known as your 20,000 × g supernatant. The pellet can be safely discarded.

4. Slowly add *ground* ammonium sulfate to the supernatant so that it becomes a 40% saturated solution. This requires 0.242 g of ammonium sulfate for every milliliter of supernatant you have. Make sure that the ammonium sulfate is mixed in slowly and completely and let the sample sit on ice for 15 min.

5. Centrifuge at 15,000 × g for 15 min at 4°C. *Save 0.5 mL of the supernatant* as the *40% supernatant*. Discard the pellet.

6. To the rest of the supernatant, add ammonium sulfate until the final concentration makes a 65% saturated solution. This requires 0.166 g of ammonium sulfate per milliliter. Let this stand for 15 min as before. Spin at 15,000 × g as before.

7. *Save 0.5 mL of the supernatant* as your *65% supernatant*. Note that this fraction should not have significant activity, but save the rest of the supernatant just to be sure.

8. To the pellet from the 65% cut, add 5–10 mL of 0.03 M bicine, pH 8.5, and dissolve the pellet well. *Save 0.2 mL as your 65% pellet.*

9. Place the remainder of the 65% redissolved pellet into a dialysis bag and suspend in a bucket of 0.03 M bicine (Q-Sepharose buffer) in a 4°C cold room.

Assaying for LDH

Each fraction listed (shown in italics) needs to be assayed today. Enzymes are notoriously unstable in crude form, and we need to find out how much activity is there before they get chopped up by proteases.

LDH will be assayed by monitoring the formation of NADH, which absorbs at 340 nm. Each assay has a reaction volume of 3 mL and contains the following:

1.9 mL of CAPS buffer, 0.14 M, pH 10

0.5 mL of NAD^+, 6 mM

0.5 mL of lactate, 0.15 M

The reaction is initiated by adding a sample composed of enzyme plus water totaling 100 μL. For your cruder samples, start with 10 μL of sample and 90 μL of water. Even 10 μL may be too much.

Whenever possible, use reaction cocktails. If you think that you are going to do ten assays, then mix up ten assays worth of ingredients into a flask:

10 × 1.9 mL CAPS   =   19 mL

10 × 0.5 mL lactate   =   5 mL

10 × 0.5 mL NAD^+ =   5 mL

Once that is made up, pipet 2.9 mL of it each time into the assay vessel. Then add 100 μL of the enzyme sample. This 100 μL may be all enzyme sample or may be some water and some enzyme sample, always totaling 100 μL. Add the extra water to the cocktail first and then add the enzyme to initiate the reaction. Invert three times and put it in the spectrophotometer. *Do not mix the enzyme and the water separately first.*

Recovery Tables

You will eventually make a recovery table for your LDH purification. For each fraction that you have (20,000 × *g* supernatant, 40% supernatant, and so on), calculate the units, units/milliliter, total units in the fraction, percent recovery, specific activity, and fold purification. It might help you to also have columns for sample volume assayed, dilution factors, and Δ Absorbance/minute.

For example, if you have 20 mL of crude homogenate, put 20 μL of it along with 80 μL of buffer into your reaction vessel, and see a Δ Absorbance/minute of 0.31, your calculations look like this:

$$U = \frac{0.31}{6220 \text{ M}^{-1}} \times 10^6 \text{ μM/M} \times 0.003 \text{ L} = 0.15$$

$$U/mL = \frac{0.15 \text{ U}}{0.02 \text{ mL}} = 7.5$$

If you make a dilution of 10 to 1 before you put the 20 μL of crude homogenate into the reaction vessel, this number is multiplied by 10:

$$\textbf{total units} = 7.50 \text{ U/mL} \times 20 \text{ mL} = 150$$

$$\textbf{\% recovery} = \frac{150 \text{ U}}{150 \text{ U}} = 100$$

By definition, the percent recovery is 100% for whatever you start with, which is the crude homogenate in this case. For the other samples, divide the total units of the sample by the total units of the crude homogenate and multiply by 100 to get percent recovery.

After doing the protein determinations on the last day, calculate the rest. If you put 20 μL of a 10-to-1 dilution of crude homogenate into the Bradford protein assay and your standard curve indicates 3 μg of protein, then your protein concentration is

$$\textbf{protein concentration} = \textbf{mg/ml}$$

$$\frac{3 \text{ μg}}{20 \text{ μL}} \times 10 = 1.5 \text{ mg/mL}$$

Specific activity is the activity per milligram of protein, which is most easily found by taking the relative activity and dividing by the protein concentration:

$$\text{specific activity} = \text{U/mg} = \frac{\text{U/mL}}{\text{mg/mL}} = \frac{7.5 \text{ U/mL}}{1.5 \text{ mg/mL}} = 5 \text{ U/mg}$$

Fold purification is the ratio of the specific activity of a given fraction divided by the specific activity of the crude homogenate.

Analysis of Results

Experiment 4a: **Purifying LDH (Comprehensive Version)**

Data

1. Provide all information on your purification steps:

   Milliliters of crude homogenate used _____

   Milliliters of 20,000 × g supernatant _____

   Grams of $(NH_4)_2SO_4$ used for 40% cut _____

   Milliliters of 40% supernatant _____

   Grams of $(NH_4)_2SO_4$ used for 65% cut _____

   Milliliters of 65% supernatant _____

   Milliliters of 65% pellet after resuspending _____

2. Provide all information on your enzyme assays:

Fraction Isolated	Quantity of Sample Assayed (μL)	Dilution Made (if any)	Absorbance Change per Minute
Crude homogenate			
20,000 × g supernatant			
40% $(NH_4)_2SO_4$ supernatant			
65% $(NH_4)_2SO_4$ pellet			
65% $(NH_4)_2SO_4$ supernatant			

Calculations

1. For each fraction, calculate the activity in the reaction vessel (units in μmol/min) and the relative activity of the sample (U/mL of fraction). The extinction coefficient is $6220\,M^{-1}\,cm^{-1}$. The cuvettes are 1 cm in diameter.

$$U = \frac{\Delta A/\Delta \min}{6220\,M^{-1}cm^{-1}\,(1\,cm)} \times 10^6\,\mu M/M \times 3 \times 10^{-3}\,L$$

$$U/mL = \frac{\text{units from above}}{\text{volume of fraction assayed}} \times \text{dilution used, if any.}$$

2. Calculate the total units for each fraction by multiplying the relative activity by the total volume of the fraction.

3. Calculate the percent recovery of the purified fractions compared to the crude homogenate.

All these calculations may be summarized in the following table.

Fraction	Units	U/mL	Total Units	% Recovery
Crude homogenate				
20,000 × g supernatant				
40% $(NH_4)_2SO_4$ supernatant				
65% $(NH_4)_2SO_4$ pellet				
65% $(NH_4)_2SO_4$ supernatant				

Questions

1. Assayed for LDH activity were 5 μL of a sample that was diluted 6 to 1. The activity in the reaction vessel, which has a volume of 3 mL, is 0.30 U. What is the ΔA/min observed? What is the relative activity of the original sample?

2. Of the original "undiluted" sample from Question 1, 10 μL of a 5-to-1 dilution were used to measure protein concentration. With the use of a standard curve, this is found to be 140 μg. What is the specific activity of the original sample?

3. If all students start with the same 10,000 × g supernatant, by the time different groups have done the 40–65% ammonium sulfate cut, their results usually vary greatly. What are likely reasons for this variance?

4. What is the difference between a percent saturated solution of ammonium sulfate and a percent w/v solution of ammonium sulfate?

5. What can you conclude about the effectiveness of these techniques for purifying LDH? How do your data support your conclusions?

6. If you see a procedure that leads to a decrease in the specific activity of LDH, what are two possible reasons?

Additional Problem Set

1. If bovine LDH is known to have a temperature optimum around 37°C, why is it important to purify it at 4°C?

2. Explain why you should not lose any LDH activity when you centrifuge a crude sample at $20,000 \times g$.

3. Explain why two different percent saturation levels of ammonium sulfate are used in these experiments.

4. Explain the physical interactions of the molecules that lead to proteins falling out of solution in high salt. What is the driving force of protein precipitation?

5. Explain why it is important to add ammonium sulfate slowly.

6. An enzyme-catalyzed reaction produces a product that has a maximum absorbance at 412 nm. The extinction coefficient is $4000 \, \text{M}^{-1} \, \text{cm}^{-1}$. A 10-to-1 dilution of the enzyme is made and 20 μL of the dilution are put into a cuvette with 80 μL of water and 1.9 mL of reaction cocktail. The absorbance change per minute is 0.05.

   a. How many units are in the cuvette if $1 \, \text{U} = 1 \, \mu\text{mol/min}$?

   b. How many units are in the cuvette if $1 \, \text{U} = 1 \, \text{nmol/min}$?

   c. What is the relative activity of the undiluted enzyme ($1 \, \text{U} = 1 \, \mu\text{mol/min}$)?

7. An enzyme sample contains 24 mg protein/mL. Of this sample, 20 μL in a standard incubation volume of 0.1 mL catalyzed the incorporation of glucose into glycogen at a rate of 1.6 nmol/min. Calculate the velocity of the reaction in terms of the following:

   a. Micromoles/minute

   b. Micromoles/liter/minute

   c. Micromoles/milligrams of protein/minute

   d. Units/milliliter

   e. Units/milligrams of protein

8. Of the sample in Problem 7, 50 mL were fractionated by ammonium sulfate precipitation. The fraction precipitating between 30 and 50% saturation was redissolved in a total volume of 10 mL and dialyzed. The solution after dialysis had 12 mL and contained 30 mg protein/mL. Of the purified fraction, 20 μL catalyzed the reaction rate of 5.9 nmol/min under the standard assay conditions. Calculate the following:

   a. The recovery of enzyme after the ammonium sulfate step

   b. The fold purification after the ammonium sulfate step

Webconnections

For a list of websites related to the material covered in this chapter, go to **Webconnections** at the *Experiments in Biochemistry* site on the Brooks/Cole Publishing website. You can access this page at http://www.brookscole .com and follow the links from the chemistry page.

References and Further Reading

Boyer, P. D., ed. *The Enzymes*, 3rd ed. 13 vols. New York: Academic Press, 1970–1976.

Boyer, R. F. *Modern Experimental Biochemistry*. Menlo Park, CA: Addison-Wesley, 1993.

Clarke, A. R., H. M. Wilks, D. A. Barstow, T. Atkinson, W. N. Chia, and J. J. Holbrook. "An Investigation into the Contribution Made by the Carboxylate Group of an Active Site Histidine–Aspartate Couple to Binding and Catalysis in Lactate Dehydrogenase." *Biochemistry* 27 (1988).

Grau, U. M., W. E. Trommer, and M. G. Rossman. "Structure of the Active Ternary Complex of Pig Heart Lactate Dehydrogenase." *Journal of Molecular Biology* 151 (1981).

Holbrook, J. J., A. Liljas, S. J. Steindel, and M. G. Rossmann. *Lactate Dehydrogenase*. Vol. 11 of *The Enzymes*, 3rd. ed., edited by P. D. Boyer. New York: Academic Press, 1975.

Kopperschlager, G., and J. Kirchberger, "Methods for the Separation of Lactate Dehydrogenases and Clinical Significance of the Enzyme." *Journal of Chromatography* 684 (1996).

Scopes, R. K. *Protein Purification: Principles and Practice.* New York: Springer-Verlag, 1994.

Sedmack, J. J., and S. F. Grossberg. "Protein Determinations." *Analytical Biochemistry* 79 (1977).

Taguchi, H., and T. Ohta. "D-Lactate Dehydrogenase Is a Member of D-Isomer-Specific 2-Hydroxyacid Dehydrogenase Family." *Journal of Biological Chemistry* 266 (1991).

Taguchi, H., and T. Ohta. "Histidine 296 Is Essential for the Catalysis in *Lactobacillus plantarum* D-Lactate Dehydrogenase." *Journal of Biological Chemistry* 268 (1994).

Chapter 5

Ion-Exchange Chromatography

TOPICS

Introduction

In this chapter, we study the charged nature of amino acids, peptides, and proteins. Amino acids have weak acid and weak base groups that give them a net positive, negative, or neutral charge, depending on their pH environment. This difference in charge can be exploited with the separation technique called ion-exchange chromatography.

5.1 Amino Acids as Weak Acids and Bases

Amino acids are both weak acids and weak bases because they contain both an amino group and a carboxyl group. When dissolved in water, they exist predominantly in their isoelectric form (no net charge):

$$H_3N^+ - CH - COO^-$$
$$|$$
$$R$$

This form is also called a zwitterion. If you add an acid, it accepts a proton, thereby acting as a base:

$$H_3N^+ - CH - COO^- + H^+ \longrightarrow H_3N^+ - CH - COOH$$
$$| \qquad\qquad\qquad\qquad\qquad |$$
$$R \qquad\qquad\qquad\qquad\qquad R$$
$$\text{base} \qquad\qquad\qquad\qquad\qquad \text{acid}$$

On the other hand, if you add a base, it acts like an acid and donates a proton:

$$H_3N^+ - CH - COO^- + OH^- \longrightarrow H_2N - CH - COO^-$$

$$\underset{\text{acid}}{\overset{|}{R}} \qquad\qquad\qquad \underset{\text{base}}{\overset{|}{R}}$$

If the R group is acidic (Glu, Asp), then it too has a carboxyl group that can donate a proton. If the R group is basic (Lys, Arg, His), then it has an amino group that can accept a proton. An amino acid is like any other weak acid or base except that there can be up to three functional groups to consider. However, only one is considered at a time because, at any given pH, usually only one of them is important.

Armed with a table of pK_a's and the Henderson–Hasselbalch equation, you can work amino acid buffer problems just the same as in Experiment 2.

PRACTICE SESSION 5.1

What forms of glutamic acid will be present at pH 4.3? The pK_a's are 2.2, 4.3, and 9.7.

$$H_3N^+ - CH - COOH$$
$$\overset{|}{CH_2}$$
$$\overset{|}{CH_2}$$
$$\overset{|}{COOH}$$

Glutamic acid

Glutamic acid has the form shown to the left at very low pH.

You really need only consider the side group carboxyl, but to convince you that this is true, let's determine the percentage of the functional groups in the A^- and HA forms:

α-COOH pK_a = 2.2

$$pH = pK_a + \log \frac{\text{base}}{\text{acid}}$$

$$4.3 = 2.2 + \log \frac{\text{base}}{\text{acid}}$$

$$2.1 = \log \frac{\text{base}}{\text{acid}}$$

$$\text{antilog } 2.1 = \frac{\text{base}}{\text{acid}} = 126$$

In other words, the ratio of the COO^- form to the COOH form is 126 to 1. In essence, all α-COOH is dissociated.

α-amino pK_a = 9.7

$$4.3 = 9.7 + \log \frac{\text{base}}{\text{acid}}$$

$$-5.4 = \log \frac{\text{base}}{\text{acid}}$$

antilog $-5.4 = \dfrac{\text{base}}{\text{acid}} = 3.9 \times 10^{-6}$, or all of it is essentially in the acid form: NH_3^+.

We didn't need to calculate the ratio of the base/acid for these two groups because the pH was so far away from the pK_a's of the groups. A good rule of thumb is disregard groups if the pH is about 2 units away from the pK_a. *Disregard* doesn't mean that the group is not important but that it is in one form or the other. With a difference of 2 units, only 1% of the functional group is in the lesser form. Table 5.1 gives the pK_a's for the functional groups of the 20 common amino acids. Figure 5.1 gives the structures of the amino acids.

TABLE 5.1 *Properties of the Common Amino Acids*

Amino Acid	3-Letter Code	1-Letter Code	pK_a α-COOH	pK_a α-NH$_3^+$	pK_a Side Chain
Alanine	Ala	A	2.34	9.69	
Arginine	Arg	R	2.34	9.69	12.48
Asparagine	Asn	N	2.02	9.82	
Aspartic acid	Asp	D	2.09	9.82	3.86
Cysteine	Cys	C	1.71	10.78	8.33
Glutamine	Gln	Q	2.17	9.13	
Glutamic acid	Glu	E	2.19	9.67	4.25
Glycine	Gly	G	2.34	9.60	
Histidine	His	H	1.82	9.17	6.0
Isoleucine	Ile	I	2.36	9.68	
Leucine	Leu	L	2.36	9.68	
Lysine	Lys	K	2.18	8.95	10.53
Methionine	Met	M	2.28	9.21	
Phenylalanine	Phe	F	1.83	9.13	
Proline	Pro	P	1.99	10.60	
Serine	Ser	S	2.21	9.15	
Threonine	Thr	T	2.63	10.43	
Tryptophan	Trp	W	2.38	9.39	
Tyrosine	Tyr	Y	2.20	9.11	10.07
Valine	Val	V	2.32	9.62	

(a) Nonpolar (hydrophobic)

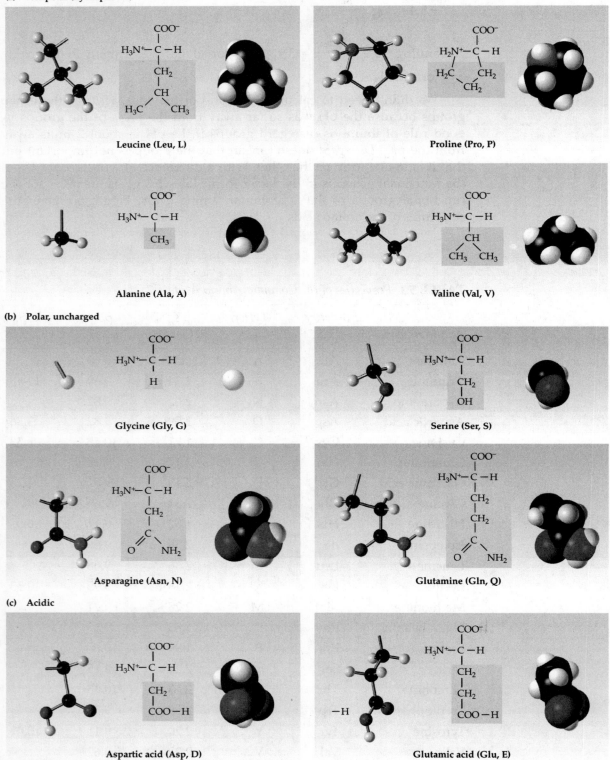

Leucine (Leu, L)

Proline (Pro, P)

Alanine (Ala, A)

Valine (Val, V)

(b) Polar, uncharged

Glycine (Gly, G)

Serine (Ser, S)

Asparagine (Asn, N)

Glutamine (Gln, Q)

(c) Acidic

Aspartic acid (Asp, D)

Glutamic acid (Glu, E)

FIGURE 5.1 *Structures of the common amino acids*

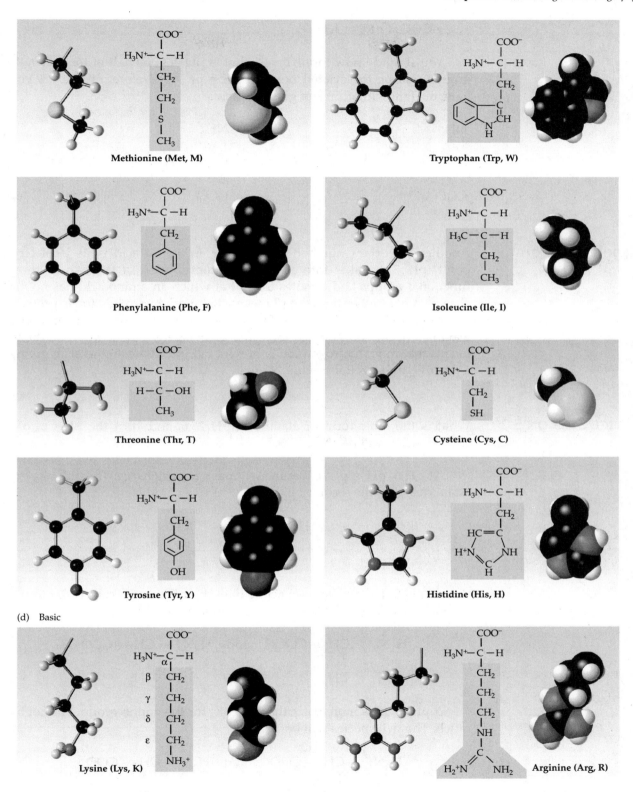

Methionine (Met, M)

Tryptophan (Trp, W)

Phenylalanine (Phe, F)

Isoleucine (Ile, I)

Threonine (Thr, T)

Cysteine (Cys, C)

Tyrosine (Tyr, Y)

Histidine (His, H)

(d) Basic

Lysine (Lys, K)

Arginine (Arg, R)

γ-COOH pK_a = 4.3

You should know immediately that at pH 4.3 one half of the γ-COOH groups will be dissociated because pH = pK_a. Therefore, at pH 4.3, you have the following forms of glutamic acid:

$$\underset{\underset{\displaystyle COO^-}{|}}{\overset{\overset{\displaystyle NH_3^+}{|}}{CH}}-CH_2-CH_2-COOH \rightleftharpoons \underset{\underset{\displaystyle COO^-}{|}}{\overset{\overset{\displaystyle NH_3^+}{|}}{CH}}-CH_2-CH_2-COO^- + H^+$$

5.2 Isoelectric Point

Another important number to know for an amino acid is its isoelectric point (pI). This takes into account the contribution of the two or three functional groups and denotes the pH at which an amino acid has no net charge. A pH higher than the pI means that more hydroxide ion is present. More hydroxide ion will pull off more hydrogen ions from the amino acid, leaving a negatively charged molecule. Conversely, a pH lower than the pI means that more hydrogen ion is present to bind to the amino acid, giving a positively charged molecule.

PRACTICE SESSION 5.2

What is the ionic form of alanine at pH 2, 6, and 10 if the pI is 6, pK_a COOH = 2.3, and pK_a NH$_3$ = 9.7?

At pH 6, pH = pI, so alanine will have no *net* charge. The only way to attain this is by the form

$$\underset{\underset{\displaystyle CH_3}{|}}{H_3N^+-CH-COO^-}$$

At pH 2, alanine is more acidic than the pK_a for the carboxyl, but not by much. Therefore, the two predominant forms will be

$$\underset{\underset{\displaystyle CH_3}{|}}{H_3N^+-CH-COOH} \quad\text{and}\quad \underset{\underset{\displaystyle CH_3}{|}}{H_3N^+-CH-COO^-}$$

A pH of 10 is more basic than the pK_a for the amino group, but not by much. The two species will be

$$\underset{\underset{\displaystyle CH_3}{|}}{H_3N^+-CH-COO^-} \quad\text{and}\quad \underset{\underset{\displaystyle CH_3}{|}}{H_2N-CH-COO^-}$$

Determining the Isoelectric Point

There is a simple way to determine the isoelectric point of an amino acid. The equation we see written is $pI = (pK_{a1} + pK_{a2})/2$, but this does not tell us which pK_a's to use. Several amino acids have three pK_a's, so you might accidentally average the wrong ones if you do not have a consistent procedure for doing this. Let's consider the amino acid lysine. The pK_a's are the following: 2.2 (carboxyl), 9.0 (α-amino), and 10.5 (side-chain amino acid). Start by drawing the amino acid with all hydrogens in place:

$$H_3N^+ - CH - COOH$$
$$|$$
$$(CH_2)_4$$
$$|$$
$$NH_3^+$$

Now, remove the hydrogen from the group with the lowest pK_a because that is where the first hydrogen comes from. Now the form looks like this:

$$H_3N^+ - CH - COO^-$$
$$|$$
$$(CH_2)_4$$
$$|$$
$$NH_{3+}$$

Does this form have zero net charge? The answer is no. Therefore, remove another hydrogen from the group with the next pK_a. Now the molecule looks like this:

$$H_2N - CH - COO^-$$
$$|$$
$$(CH_2)_4$$
$$|$$
$$NH_{3+}$$

Does this form have zero net charge? The answer is yes. To get the pI, average the pK_a of the last group where the hydrogen was removed with the first pK_a that did not have a hydrogen removed. In this case, average 9.0 and 10.5, which gives a pI of 9.8.

PRACTICE SESSION 5.3

What is the pI for a mythical amino acid that has two carboxylic acid groups (one with a pK_a of 2 and one with a pK_a of 4) and two amino groups (one with a pK_a of 8 and one with 10).

Without even trying to draw such a molecule, we can still calculate the pI. Starting with the molecule with all hydrogens in place, the molecule has a net charge of +2. Removing the hydrogen from the most acidic group (pK_a = 2), we have one COO^- group, so the net charge is +1. Removing the next hydrogen (the pK_a = 4), we have another COO^- group, so the net charge is 0. At that point, average 4, which is the pK_a for the last group where a hydrogen was removed, with 8, which is the next pK_a in line. The pI is therefore $(8 + 4)/2 = 6$.

This technique for determining pI also works for peptides and proteins. Just remember that the majority of the functional groups are tied up in the peptide backbone. Therefore, they do not gain or lose hydrogens and are not considered. If a peptide has ten amino acids, there will be one α-carboxyl at the C terminus and one α-amino at the N terminus. The only other groups that count are the side-chain acid or base groups.   ●

PRACTICE SESSION 5.4

Calculate the pI for the following peptide:

Phe-Lys-Glu-Asp-Lys-Ser-Ala

First, redraw this peptide, showing clearly what acid or base groups are present. The α-amino group is on Phe, a side-chain amino group is on Lys, a side-chain carboxyl is on the Glu, a side-chain carboxyl is on Asp, another side-chain amino is on the next Lys, nothing is on the serine, and the α-carboxyl group is on Ala:

$$H_3N^+ - Phe - Lys - Glu - Asp - Lys - Ser - Ala - COOH$$
$$\qquad\qquad\quad | \qquad\quad | \qquad\quad | \qquad\quad |$$
$$\qquad\qquad NH_3^+ \quad COOH \ COOH \ NH_3^+$$

Second, determine the pK_a's for the various groups shown (see Table 5.1). From left to right, they are 9.13, 10.5, 4.3, 3.9, 10.5, and 2.3.

Third, remove hydrogens from the various groups starting with the lowest pK_a until the form with no net charge is present. The preceding peptide shows three positive charges. Thus, to attain the isoelectric form, remove hydrogens from three groups that will give negative charges. The peptide now looks like this:

$$H_3N^+ - Phe - Lys - Glu - Asp - Lys - Ser - Ala - COO^-$$
$$\qquad\qquad\quad | \qquad\quad | \qquad\quad | \qquad\quad |$$
$$\qquad\qquad NH_3^+ \quad COO^- \ COO^- \ NH_3^+$$

Finally, average the pK_a from the last group where a hydrogen was removed (the one with the highest pK_a of those where a hydrogen was removed) with the pK_a of the next group in line (the one with the lowest pK_a of those where a hydrogen was not removed):

$$pI = \frac{4.3 + 9.0}{2} = \textbf{6.65}$$

There are many ways of calculating pI's, but students who learn this method rarely make a mistake!

5.3 Ion-Exchange Chromatography

One of the most effective and most used methods for separating charged compounds is **ion-exchange chromatography,** often abbreviated IEX. An ion-exchange column is a column full of a resin that contains charged groups on the surface. We will use a type of ion-exchange resin called Dowex 50, which has a sulfonic acid group as shown in Figure 5.2.

Originally, the resin is in the acid-washed, or H^+, form. Then the resin is reacted with NaOH to strip off the dissociable hydrogens, giving the sodium form. It is this sodium form that is the ion exchanger because the sodium ions can be **exchanged** for other positively charged molecules. That is, a positively charged molecule, such as an amino acid below its pI, could bind to one of the sulfonyl groups displacing a sodium ion. Because it is a cation that binds and exchanges, this type of column is also called a **cation-exchange column.**

By running a mixture of amino acids over an ion-exchange column, you can separate the amino acids. Figure 5.3 demonstrates this principle. Let's assume that we have the Dowex column already in the sodium form. Then take a mixture of three amino acids (histidine, serine, and aspartic acid) and run the mixture through the column. Everything is in a pH 3.25 buffer. At that pH, the aspartic acid has two forms: 80% is in a form with no net charge, and 20% is in a form with a net negative charge. This means that the aspartic acid will not bind to the ion exchanger and so elutes quickly. The serine is electrically neutral, so it also passes through the column

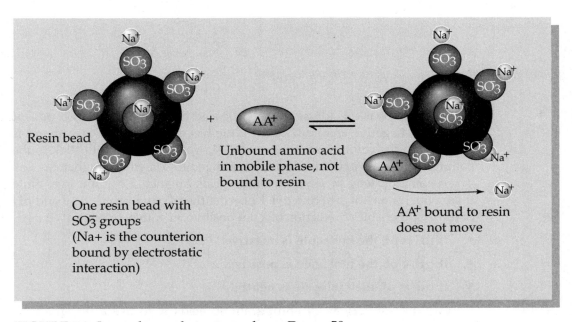

FIGURE 5.2 *Ion-exchange chromatography on Dowex 50*

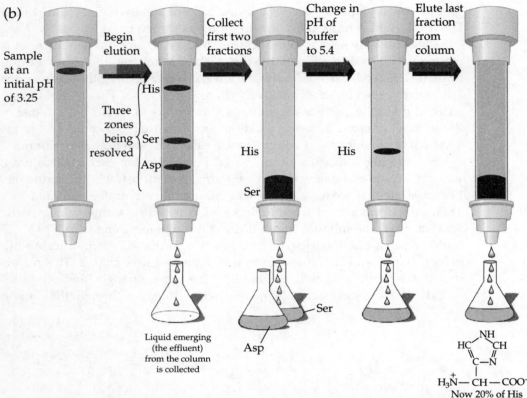

FIGURE 5.3 *Elution of amino acids with ion-exchange chromatography*

quickly but not as quickly as the aspartic acid because it has no component with a net negative charge. The histidine has a net positive charge and is attracted to the column, so it moves the slowest. In fact, it will stick to the column unless the pH of the buffer is changed to something higher.

A common way to separate amino acids on such a column is to elute them with buffers of differing pH. Knowing the pI of the amino acid and pH of the buffer enables prediction of a net positive, negative, or neutral charge:

- **If pH > pI, the molecule is negative.**

- **If pH < pI, the molecule is positive.**

- **If pH = pI, the molecule is neutral.**

A more subtle way of eluting amino acids is to use a pH gradient. Instead of using different buffers of different pHs, a gradient maker can be used to create a buffer of continually increasing pH. In this way, amino acids with small differences in pI can also be separated. Sometimes it is undesirable to change the pH of the buffers because doing so may damage the molecules that you are separating. It is also possible to elute amino acids or other charged molecules by adding excesses of the same charge. For instance, by adding a high concentration of sodium ions to the Dowex column described earlier, amino acids could be eluted as the sodium ion outcompetes them for the sulfonyl sites.

PRACTICE SESSION 5.5

A mixture of Phe (pI 5.5), Thr (pI 6.5), Asp (pI 3), His (pI 7.6) and Lys (pI 9.8) at pH 11 is loaded on an anion-exchange column and then eluted with a pH gradient of lowering pH. What order will the amino acids elute?   ●

At pH 11, all will be negatively charged because all pI's are less than 11. Amino acids will elute in order as their pI's are reached, so the order will be Lys, His, Thr, Phe, and Asp.

5.4 Ion-Exchange Resins

Ion-exchange resins are made up of two parts. The first is the insoluble, three-dimensional matrix or support. The second is the chemically bonded charged group that is responsible for the ion-exchange effect. The gel matrix can be made from a variety of materials. Common ones are polystyrene resins, like Dowex; agarose resins, which carry the common trade name Sepharose; dextran resins, which are usually sold as Sephadex; and polyacrylamide resins, which are usually called Bio Gel. All can have controlled pore sizes that affect the ability of a molecule to enter the inside of the gel matrix. For small molecules like amino acids or small proteins, polystyrenes like Dowex can be used. For larger molecules, like most enzymes, a larger pore size is needed, so dextrans and agarose-based gels are used.

The most important part of an ion exchanger is the charged group that is chemically bonded to the support. A **cation exchanger** is a negatively charged group such as a sulfonic acid or carboxylic acid. An **anion exchanger** is a positively charged group such as a substituted amino group. Ion exchangers are also classified based on the ionizing strength of the charged group. A quaternary amine is always charged, as is a sulfonic acid. These are a strongly basic and strongly acidic ion exchanger, respectively. Other groups ionize to a lesser extent such as the weak anion exchanger, DEAE-Sephadex, and the weak cation exchanger, CM-Sephadex. Table 5.2 lists some typical-ion exchange resins.

When selecting an appropriate ion-exchange resin, first decide whether you need a cation or an anion exchanger. This depends on what you are trying to purify and the pH that you need to use in the column. If you select an exchange column, then you also need to decide if you want the functional group to be strong or weak. Strong exchangers like Dowex are usually reserved for sturdier materials such as amino acids. Many enzymes will be denatured in the environment of a strong exchanger.

TABLE 5.2 *Typical Ion-Exchange Resins*

	Functional Group	*Matrix*	*Class*
Anion Exchangers			
AG 3	Tertiary amine	Polystyrene	Weak
DEAE-cellulose	Diethylaminoethyl	Cellulose	Weak
DEAE-Sephadex	Diethylaminoethyl	Dextran	Weak
QAE-Sephadex	Diethyl-(2-hydroxyl-propyl)-aminoethyl	Dextran	Strong
Q-Sepharose	Tetramethyl amine	Agarose	Strong
Cation Exchangers			
Dowex	Sulfonic acid	Polystyrene	Strong
CM-cellulose	Carboxymethyl	Cellulose	Weak

5.5 Identification of Eluted Compounds

When performing an IEX, you put a sample of biological molecules on the column. Some will stick to the column while others wash off. Then elute the bound molecules with a salt gradient or a pH change. Either way, now find the eluted molecules and identify them. Very often, spectrophotometry with or without a colorimetric assay is used to find the column fractions that contain the molecules of interest.

Detection of Amino Acids

If some amino acids are eluted from a column, how do we find them? They are colorless and only a few can be measured spectrophotometrically. To do this, we usually react the amino acids with ninhydrin, which gives a purple color with all of the amino acids except proline. Figure 5.4 shows the reaction.

Identification of Proteins

When proteins are separated with IEX, fractions are obtained by collecting drops off the column into different test tubes. Some tubes will have the proteins of interest in them. Usually, a UV monitor is attached to the column outlet so that the absorbance at 280 nm can be measured while the column is running. When the absorbance is high, protein is eluting from the column. This tells which tubes had protein in them. In a mixture of proteins, however, the desired protein may not have been isolated. The UV monitor helps narrow down the choices. Instead of searching for the desired protein in 200 samples, your search may now be limited to 20 or 30. Then assay the likely fractions for the protein by whatever specific assay is appropriate for that protein. When a flow-through UV monitor is not available, assay fractions at 280 nm with a standard UV spectrophotometer. If this is also not available, then assay each fraction for protein, using a colorimetric assay, or for the specific enzyme via its enzymatic assay.

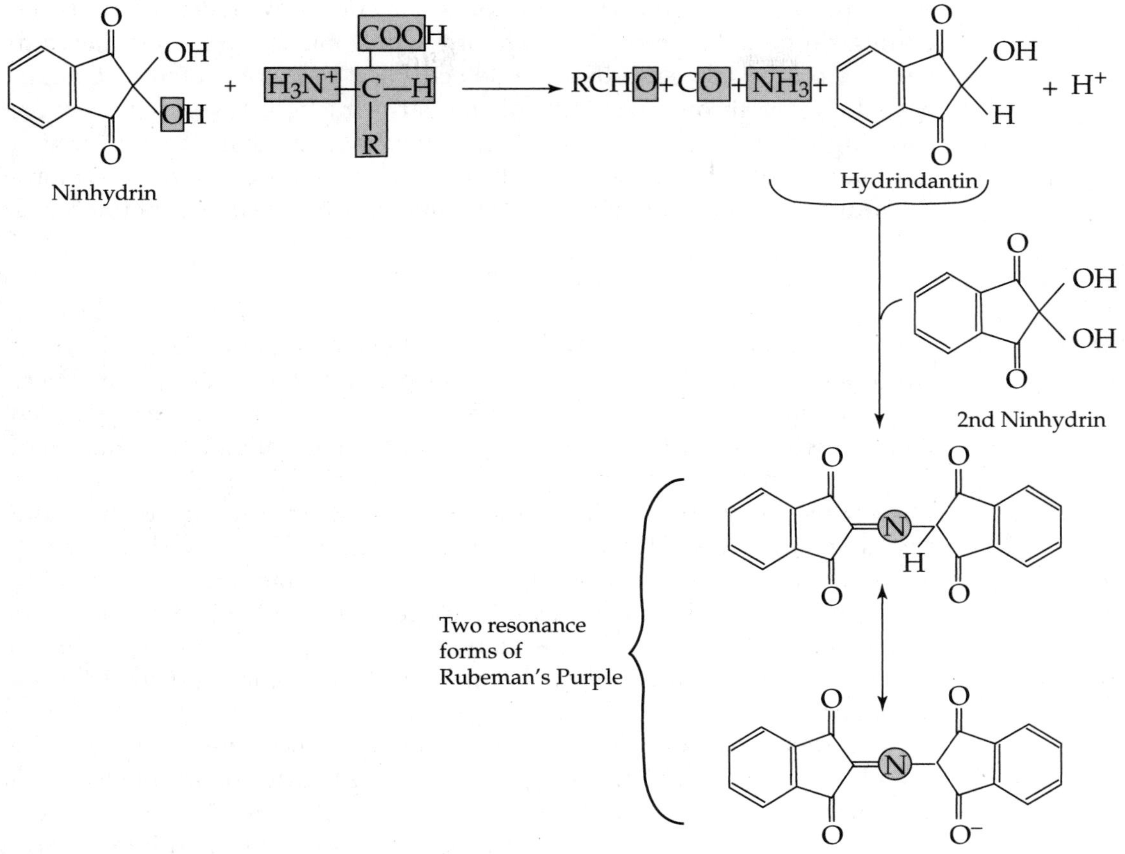

FIGURE 5.4 *The ninhydrin reaction with amino acids*

Column Attachments

A column chromatography setup can be very simple, just the resin and a column, or can be very complex with a column, fraction collector, pump system, and UV monitor. Indeed, some systems take longer to set up than to run the experiment. When using such attachments, make sure that *all* tubing is connected correctly and tested. Let the column run for a few minutes before loading your sample. This ensures that nothing is leaking and the detector and collector are working.

Another concern when using column attachments is the calibration of the equipment. Fraction collectors can be set to change tubes based on time or by the number of drops added. The pumps may indicate that a certain fraction volume will be obtained with a particular setting, but it is always safest to check this. While running the column to make sure that the lines are working, divert the solution to a fraction collector to measure the actual volume produced per time or per drop. Then set the fraction collector to give the fraction size wanted.

You may also use a UV monitor to let you know when proteins are eluting from a column. As with any electronic device, "Garbage in, garbage out." Make sure that the detector and monitor are correctly calibrated. Never throw away any of the early fractions, even if they were showing no absorbance, until you find the fractions that you are looking for. Many people have discarded the early fractions because the UV monitor indicated no absorbance, only to later find out that the monitor was not calibrated.

5.6 Thin-Layer Chromatography

Another technique often used to identify specific amino acids is **thin-layer chromatography** (TLC). TLC is a type of partition chromatography. There are two phases: stationary and mobile. Molecules are partitioned between the two, with their affinity for one or the other controlled by the nature of the stationary-phase matrix and the mobile-phase buffer. Amino acid unknowns are spotted on a thin-layer chromatogram along with amino acid standards. The TLC is then developed in a solvent, and the amino acids travel up the chromatogram at characteristic rates dependent mainly on the polarity of the amino acid. Silica gel is very polar, so polar amino acids will tend to stay with the stationary silica phase. Nonpolar amino acids will tend to travel with the mobile phase, which is usually more nonpolar.

The quality of your TLC is based primarily on how well you apply the samples. Very small sample spots must be applied to the thin-layer resin because they will spread as the chromatogram develops. The quality of the chromatogram plate and the care that is used when putting it into the solvent are also very important.

Rf Values

Have you ever wondered why we calculate Rf's for many types of chromatography? Whether it is called an Rf, as in this experiment, or an Rm, as in many types of gel electrophoresis, it is common to divide the distance a sample migrated by the distance of a solvent front of some kind. There are two reasons for doing this:

1. *Nonlinear solvent front.* If the solvent front is not straight, such as in Figure 5.5, two spots that ran the same distance may in fact not be the same compound. Because the solvent front ran differently, samples appear to be the same but are not. Calculating an Rf will give different numbers for these two samples.

2. *Reproducibility between experimenters.* If we run these amino acids in the same solvent on the same thin-layer support, we may or may not get the same distances. If we let our samples run half as long as yours do, our numbers will not be comparable. However, by calculating Rf's, the ratios should be very similar.

FIGURE 5.5 *Thin-layer chromatography with an irregular solvent front*

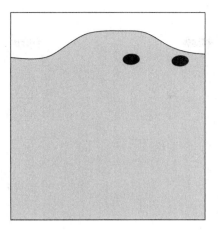

5.7 Why Is This Important?

The two techniques we will learn are very important to most of the life sciences. IEX is used in every subfield of biochemistry and related sciences. By altering the exchange resin, we can control the types of molecules that bind. We can purify a component by binding the impurities or by binding the component of interest and letting the impurities wash through. Two different columns can be used that will accomplish both. A popular separation technique that allows several components to be resolved in a few minutes is high-performance liquid chromatography (HPLC). Many of the columns used with HPLC are just fancy ion-exchange columns. When you take away all of the extra peripherals, you get back to basic IEX. This type of chromatography is commonly used with DNA purification. The costly and somewhat dangerous technique of cesium chloride centrifugation has largely been replaced by small ion-exchange columns. Plasmid DNA preps used to take hours but now can be done in minutes with the Qiagen ion-exchange system.

TLC is equally useful. As you will see, good separations can be attained in a short time. Different matrices exist for the separation of all kinds of biomolecules. Many complicated detection systems employ TLC; drug analysis is just one of them. The preliminary screen of an athlete's urine sample is usually done with a TLC plate.

Experiment 5

Separating and Identifying Amino Acids

In this experiment, you will conduct an analysis of a solution containing several amino acids. The amino acids will be separated by cation-exchange chromatography and then identified by TLC. By planning ahead and working efficiently together, you and your lab partner can be done on time.

Prelab Questions

1. In Part B, step 3, you are to test the effectiveness and specificity of ninhydrin by spraying spots of water, the protein bovine serum albumin (BSA), and amino acids with ninhydrin. What do you predict the results will be, and why? What assumptions must you make?

2. Summarize in a couple of sentences what you are going to do in this experiment, why you are using two different techniques, and the information that you will get from each.

Objectives

Upon successful completion of this lab, you will be able to

- Predict the dominant ionic form of an amino acid at a given pH.
- Calculate the percentage of the amino acid in the various forms at a given pH.
- Explain how anion- and cation-exchange columns work.
- Predict the order of elution of amino acids off an ion-exchange column.
- Separate a mixture of amino acids into acidic, basic, and neutral amino acids using a Dowex 50 column.
- Apply samples properly to columns and thin-layer chromatograms.
- Identify amino acids by TLC, using suitable standards.

Experimental Procedures

Materials

Bovine serum albumin

Dowex 50 × 8 (200–400 mesh) in 0.05 M citrate, pH 3

Citrate buffer, 0.05 M, pH 3 and 6

CAPS buffer, 0.05 M, pH 11

Glass columns

Amino acid mixtures, 1% in pH 3 citrate (possibilities will be on the chalkboard)

Ninhydrin solution

Whatman 3 MM chromatography paper, 3 × 10 cm

TLC plates, 7 × 10 cm

Propanol/acetic acid/water (PAW) solvent, 4/1/1

Drawn-out capillary tubes for spotting paper chromatograms

Amino acid standards, 2% w/v in water

Procedures

Part A: Separating and Identifying Amino Acids by TLC

1. Obtain a 7- × 10-cm thin-layer chromatogram. Handle only with gloves and do not touch the gel surface. The chromatogram is prepared by spotting along one edge.

2. Draw a line on a piece of paper. Set the chromatogram gel side up on the paper, such that the line is 1.5 cm from the bottom. Do *not* write on the chromatogram itself.

3. Make room for seven spots: six amino acid standards and one unknown. Prepare a key to indicate where the amino acids are located on your paper.

4. Using drawn-out capillary tubes, spot one application of the amino acids. Make the spots as small as possible (1–2 mm).

5. After all spots have dried, carefully place the chromatogram into a beaker with 1 cm of PAW solvent in the bottom. Put the chromatogram in straight, with the samples at the bottom. The solvent must be below the line of samples. Cover the beaker with aluminum foil. Proceed to Part B, while the chromatogram is running.

6. Let the solvent proceed until three fourths of the way to the top. Use a pencil to mark where the solvent front was when you took it out of the solvent.

7. Spray the chromatogram with ninhydrin in a fume hood (*wear gloves*) and develop in an oven (110°C) for 10 minutes or carefully dry with a hair dryer.

Chromatogram after marking the solvent front and spraying with ninhydrin

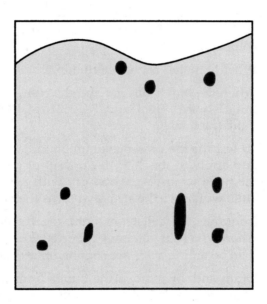

Part B: Separating Amino Acid Mixtures by Ion Exchange

1. Obtain a Dowex 50 column. These have been prepared in the sodium form and are equilibrated in 0.05 M citrate buffer, pH 3.

2. To check the ninhydrin reaction, take one strip of filter paper and place one spot of water, one spot of your original unknown, and one spot of the protein BSA. Spray with ninhydrin, heat in an oven or with a hair dryer, and compare the color and intensity.

3. Remove the plug on the column and allow the buffer on top of the column to drain until the meniscus is just above the resin and then replug.

4. From now on, *do not allow the column to run dry!* Always use a Pasteur pipet to add solution to the column.

5. Prepare 30 or so test tubes for fraction collecting by marking with a line at the 1-mL level.

6. Carefully apply 0.5 mL of amino acid unknown to the top of the resin. Do not disturb the resin.

7. Collect drops into tube 1.

8. When the meniscus has just entered the resin, replug the column, and add 1 mL of 0.05 M citrate buffer, pH 3, carefully. Wash the sides of the column, continue collecting.

9. When 1 mL has been collected in tube 1, change to tube 2.

10. When the buffer has entered the resin, repeat the addition of 1 mL of buffer twice. Then add 5–10 mL of buffer and continue collecting 1-mL fractions.

11. To test for the presence of an amino acid, take 1 drop of the fraction and apply to the 3- × 10-cm strip of paper. Spray with ninhydrin in the hood (*wear gloves*) and dry with a hair dryer. For each buffer you elute with, save the strip that turns the darkest purple.

12. Continue the collection until the fractions no longer turn purple. Alternatively, if you have collected 20 mL and still see no purple color, then no amino acids are coming off at that pH.

13. Drain and/or remove the buffer from the top and replace carefully with 1 mL of 0.05 M citrate buffer, pH 6. Repeat steps 8–12 with this buffer until again no more amino acids come off.

14. Repeat with 0.05 M CAPS, pH 11.

Name _____   Section _____

Lab partner(s) _____   Date _____

Analysis of Results

Experiment 5: **Separating and Identifying Amino Acids**

Data   Unknown _____

Part A

1. Draw a picture of your TLC results in the box. You may also attach your thin-layer chromatogram if so directed.

2. Label the drawing so that it is clear which amino acids are in each column.

3. Calculate the Rf for each spot on your chromatogram. The Rf is the ratio of the distance the spot moved to the distance the solvent moved.

Amino Acid	Rf
_____	_____
_____	_____
_____	_____
_____	_____
_____	_____
_____	_____
Unknown	_____

Part B

1. Of the three buffers that you used, which ones caused the elution of amino acids?

2. Attach your paper strips to this report as justification for your answer to Question 1.

Analysis and Questions

1. Where should the amino acids that you used elute from the column— that is, with which buffers?

Amino Acid	Buffer It Should Elute With
_____	_____
_____	_____
_____	_____
_____	_____
_____	_____
_____	_____

2. Based solely on the information from the ion-exchange column, which amino acids can you have in your unknown?

3. Identify the amino acids in your unknown. If you cannot decide between possibilities, at least narrow them down.

4. Determine the isoelectric point of the following peptide:

$$H_3N^+ — Glu — Lys — Leu — Asp — Glu — His — COOH$$

Show your work on how you calculated your answer.

5. Draw all forms of histidine at pH 6.5 present at a level of at least 10% of the total possible forms.

6. Why shouldn't ninhydrin react with a protein?

7. How would the results have been different for the ion-exchange part of the experiment if an unknown sample had glutamic acid, histidine, and arginine?

Experiment 5a

Purifying LDH with Ion-Exchange Chromatography

In this experiment, you will use ion-exchange chromatography (IEX) with Q-Sepharose to continue your purification of LDH that you started in Experiment 4a.

Prelab Questions

1. What is the chemical nature of Q-Sepharose that allows it to be used as an ion exchanger?

2. What is the relationship between the pI of LDH and the pH of the buffers that we will be using?

3. List three ways of eluting LDH from the column once it is bound.

Objectives

Upon successful completion of this lab, you will be able to

- Properly pour an ion-exchange column.
- Properly set up the column, fraction collectors, and UV monitors.
- Properly load the LDH sample on the column and elute with appropriate buffers.
- Explain how anion- and cation-exchange columns work.
- Analyze UV monitor results and predict the location of LDH.
- Assay fractions and determine recovery from column.

Experimental Procedures

Preparation of IEX Columns

Materials

Bio Rad Econo columns (1.5×15 cm) or other columns of a similar size

Fast Flow Q-Sepharose equilibrated in 0.03 M bicine, pH 8.5

Plugs, caps, tubing, and connectors for columns

Column peripherals (pumps, UV monitor, fraction collector, detector)

Procedures

1. Acquire some Q-Sepharose. This needs to be de-gassed before using.

2. Acquire a 15- $\times$ 1.5-cm column and column support.

3. Pour and pack the column until the packed resin is a few centimeters below the top.

4. Make sure that the column is well stoppered at the bottom and top. Use Parafilm on both the bottom and top to help keep it from leaking.

5. Place the columns in a designated location until ready for use.

IEX

Materials

Q-Sepharose columns

Fraction collectors

UV monitors and chart recorders

CAPS buffer, 0.15 M, pH 10

NAD^+, 6 mM

Lactate, 0.15 M

Bicine, 0.03 M, pH 8.5 (wash buffer)

Bicine, 0.03 M, pH 8.5, with 0.2, 0.4, 0.6, 0.8, or 1 M NaCl added (elution buffer)

Sodium phosphate, 0.02 M, pH 7.0 (dialysis buffer for Experiment 6.9)

Procedures

1. Collect your ion-exchange columns that you prepared previously. Collect your dialysis bags from Experiment 4a, keeping them cool until needed.

2. If you have a visible precipitate in the dialysis bag, transfer the solution to a centrifuge tube and spin for a couple of minutes, using a benchtop clinical centrifuge. Save the supernatant.

3. Save a small aliquot of the dialyzed LDH for assay. It will be very important to know how many units you are loading on the column. This assay should be done early in the day. In addition, this will tell you if you lost any activity during the dialysis.

4. Make sure that the column, fraction collector, and UV monitor are all assembled properly, turned on, and warmed up. If any of these are not available, the following procedure will have to be modified.

Ion-Exchange Loading

5. Drain the excess buffer off the top of the column by allowing the column to run with the fraction collector tube dripping into a graduated cylinder. Set the fraction collector to collect drops and set the counter to its maximum value so that it will not advance during this procedure. Figure out how many drops it takes to give you 1 mL. This is best done by collecting 5–10 mL into the graduated cylinder so that you get a better average.

6. When the resin is just about to run dry at the top, stop the column. Have the fraction collector ready to collect drops and set the volume to 3 mL.

7. Slowly load the dialyzed LDH on the column, being careful not to disturb the bed. After loading, allow the column to flow. Be very careful that it doesn't run dry. Once the last of the LDH is about to disappear into the resin, stop the column.

Ion-Exchange Wash

8. Add 0.03 M bicine, pH 8.5, and resume collecting. *Make sure that the column never runs dry.*

9. Collect fractions until the UV monitor spikes and then goes back to baseline. Mark the tubes periodically so that you know what tubes are being collected at various positions on the chart recorder. If a UV monitor is not available, you may have to assay every fraction for LDH until you find it.

10. Assay the fractions for LDH if they have significant protein as indicated by the UV monitor.

11. If any of these fractions have significant LDH activity, calculate how many units have been collected.

12. Do not throw away any fractions until you have found the LDH that you are seeking. To do this, you must know how many units you loaded on the column and how many you have found in the fractions.

Ion-Exchange Elution

13. If it appears that most of your LDH bound to the column, then you will need to elute it. Do this with a batchwise application or a salt gradient.

14. When the last of the wash buffer is disappearing into the resin, switch over to elution buffer. Start with 15 mL of bicine plus 0.2 M NaCl. If a gradient maker is available, set it up going from plain bicine to bicine plus 1 M NaCl, 40 mL each per chamber. Step 16 would then be irrelevant.

15. Collect fractions as before, watching the UV monitor carefully for evidence of peaks eluting. Hopefully, the LDH will elute with a peak because that makes it easier to find.

16. When the 0.2 M NaCl elution buffer runs out, switch to the 0.4 M NaCl buffer and so on.

17. Find the LDH and pool the fractions with the most significant activity. Note that, if you pool fewer fractions, you will have less material to work with, but it may be more pure. There are still two columns to go to attain reasonable purity, so it is best to optimize for yield rather than purity at this step.

18. Assay the pooled fractions and save some for protein determination as always.

19. Place the pooled fractions into a dialysis bag and put them into a bucket of 0.02 M sodium phosphate, pH 7.0, that is at 4°C.

Analysis of Results

Experiment 5a: **Purifying LDH with Ion-Exchange Chromatography**

Data

1. Fill in the following table concerning the dialyzed sample that you loaded on the column.

Data from Dialyzed Sample

Microliters assayed	
Dilution used (if any)	
$\Delta A / \Delta$ min	
U/mL	
Total milliliters loaded on column	
Total units loaded on column	

2. Fill in the following table for the wash buffer fractions.

Data from Wash Fractions

Fraction No.	Volume Assayed	$\Delta A/\Delta min$	U/mL	Total Units

Data from Wash Fractions

3. Fill in the following table for the elution fractions.

Data for Elution Fractions

Fraction No.	Volume Assayed	Δ A/Δ min	U/mL	Total Units

Analysis of Results

1. Fill in the following table concerning your IEX results.

Summary of Ion-Exchange Results

Fractions pooled	
Volume assayed	
$\Delta A / \Delta$ min	
U/mL	
Total units	
% recovery off of column	
% recovery from beginning of experiment	

2. What percentage of the loaded activity eluted in the wash buffer?

3. What percentage eluted in the elution buffers?

4. Which salt concentration did the most activity elute with? If you used a gradient, estimate the salt concentration from the volume, assuming a linear change from 0 to 1 M.

5. Is there a significant difference between the relative activity of the 65% pellet from Experiment 4a and the dialyzed pellet you used today? If so, how much activity did you lose? How can this be improved?

Additional Problem Set

1. Write equations to show the ionic dissociations of the following amino acids: glutamic acid, isoleucine, histidine, and lysine.

2. Predict the most prevalent form of the following amino acids at pH 7: glutamic acid, isoleucine, histidine, arginine, aspartic acid, and cysteine.

3. Based on the information in Table 5.1, are there any amino acids that make good buffers at pH 8.0? Which ones? Which one will be the best buffer?

4. Given the following peptide

Val — Met — Ser — Gly — Glu — Ser — Asp — His — Lys — Cys — Tyr — Leu

   a. What is the pI?

   b. What is the net charge at pH 7.0?

   c. What is the net charge at pH 4.0?

5. Consider the following peptides:

Phe — Glu — Ser — Met   and   Val — Trp — Cys — Leu

Do these peptides have different net charges at pH 1.0? At pH 7.0? Indicate the charges at both pH values.

6. A solution containing aspartic acid, glycine, threonine, leucine, and lysine is applied to a Dowex 50 cation-exchange column at pH 3.0. If the amino acids are eluted with an increasing pH gradient, in what order will they elute?

7. To determine the pI of an amino acid or a peptide, why is it important to know both the structure and the pK_a of the functional group instead of just the pK_a?

8. Why must the pI be an average of two pK_a's?

9. What is the most useful definition of acid and base when discussing the side chains of amino acids and their pI's?

10. What are the three amino acids that are considered basic according to the definition in Problem 9?

Webconnections

For a list of websites related to the material covered in this chapter, go to **Webconnections** at the *Experiments in Biochemistry* site on the Brooks/Cole Publishing website. You can access this page at http://www.brookscole.com and follow the links from the chemistry page.

References and Further Reading

Ahern, H. "Chromatography." *The Scientist* 10, no. 5 (1996).

Boyer, R. F. *Modern Experimental Biochemistry*. Menlo Park, CA: Addison-Wesley, 1993.

Campbell, M., and S. Farrell. *Biochemistry*. Pacific Grove, CA: Brooks/Cole, 2002.

Cooper, T. G. *The Tools of Biochemistry*. New York: Wiley Interscience, 1977.

Hamilton, P. B. "Ion Exchange Chromatography of Amino Acids: A Single Column, High Resolving, Fully Automatic Procedure." *Analytical Chemistry* 35, no. 13 (1963).

Robyt, J. F., and B. J. White. *Biochemical Techniques*. Long Grove, IL: Waveland Press, 1990.

Segel, I. H. *Biochemical Calculations*. New York: Wiley Interscience, 1976.

Shihamoto, T. *Chromatographic Analysis of Environmental and Food Toxicants*. New York: Marcel Dekker, 1998.

Tsao, G. T. *Chromatography*. San Diego: Elsevier Science, 1991.

Chapter 6

Affinity Chromatography

TOPICS

Introduction

In this chapter, we discuss another type of adsorption chromatography, called affinity chromatography, which is based on the interactions between a protein of interest and a ligand bound to the gel matrix. Because the binding of the proteins is specific to the structure of a bound ligand, excellent purifications can be achieved with this technique.

6.1 Affinity Chromatography

Affinity chromatography is a type of adsorption chromatography, similar to ion exchange. With an ion exchanger, however, the basis of the separation of molecules is the charge nature of the protein compared to the gel. With a cation exchanger, for example, any protein with a net positive charge is expected to bind. With affinity chromatography, the nature of the binding is more specific. We design or buy a column resin that has a ligand on it that our molecule recognizes. For the separation of LDH, for example, an affinity resin might resemble one of the substrates, either lactate, pyruvate, NAD^+, or NADH. Figure 6.1 demonstrates the basic principles.

A mixture of proteins is passed over the affinity column. The affinity ligand has a binding affinity for a protein of interest, which binds to the ligand. The other proteins elute from the column into the wash. Then the bound protein of interest is eluted.

Affinity chromatography can give extensive purifications in a single step because of the specific nature of the binding. In theory, you could achieve complete separation with this step if the binding between the ligand and protein were specific enough. This rarely happens in practice, of course.

6.2 Gel Supports

The gel supports used in affinity chromatography are the same ones discussed in Chapter 5. Most affinity gels are created by covalently linking the ligand to a matrix of agarose, dextran, or polyacrylamide. In the experiment

FIGURE 6.1 *The basics of affinity chromatography*

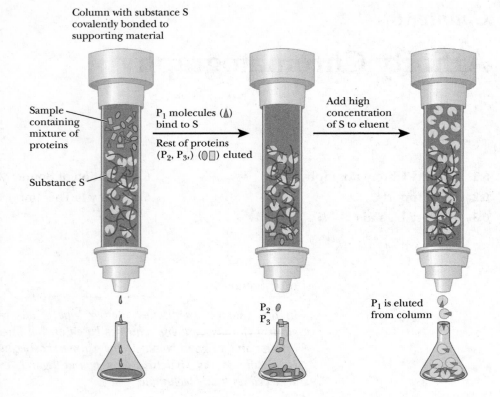

Column with substance S covalently bonded to supporting material

Sample containing mixture of proteins

Substance S

P_1 molecules (▲) bind to S

Rest of proteins (P_2, P_3,) (◐□) eluted

Add high concentration of S to eluent

P_2
P_3

P_1 is eluted from column

that follows, we use Cibacron Blue Sepharose. Sepharose is the support; Cibacron Blue is the ligand.

6.3 Affinity Ligands

The ligands used for affinity chromatography can be broken down into two broad categories: individual and group specific.

Individual Ligands

An individual ligand is very specific for a protein of interest. For example, Figure 6.2 shows the structure of NAD^+. If we make or buy an affinity resin that is composed of NAD^+ attached to a Sephadex matrix, we have a very specific affinity resin for LDH because NAD^+ is one substrate for the enzyme. The binding is expected to be very good between the ligand and the enzyme, although it is not absolutely specific because other enzymes use NAD^+ as substrates.

Group-Specific Ligands

A group-specific ligand is a little less specific in its interaction than an individual ligand. A good example is AMP–agarose. AMP has the structure shown in Figure 6.3. Look at AMP carefully and note that it is essentially

FIGURE 6.2 *Structure of NAD$^+$*

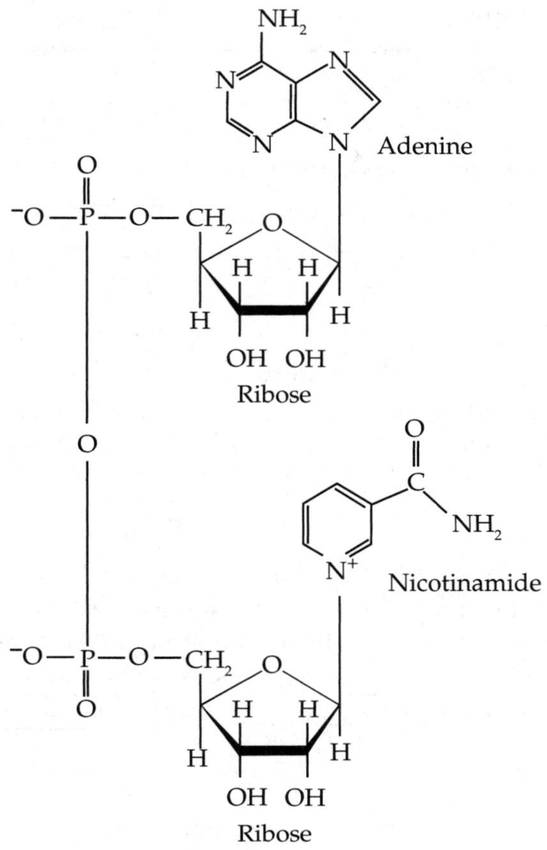

one half of NAD$^+$. AMP–agarose is expected to bind enzymes that use AMP directly or enzymes that bind NAD$^+$ cofactors because AMP fits nicely into the active site of any molecule that uses NAD$^+$. Table 6.1 gives some examples of group-specific resins.

½ NAD$^+$

FIGURE 6.3 *Structure of AMP*

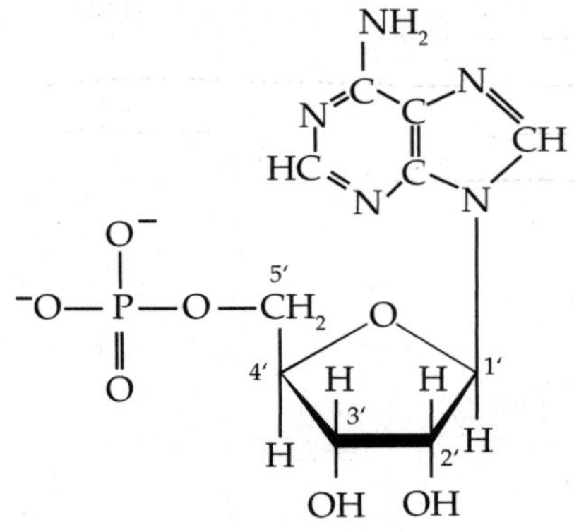

Adenosine 5′ monophosphate

TABLE 6.1 *Group-Specific Affinity Resins*

Group-Specific Adsorbent	Group Specificity
Concanavalin A–agarose	Glycoproteins and glycolipids
Cibacron Blue–agarose	Enzymes with nucleotide cofactors
Boronic acid–agarose	Compounds with *cis-diol* groups
Protein A–agarose	IgG-type antibodies
Poly(U)–agarose	Nucleic acids containing poly(A) sequences
Poly(A)–agarose	Nucleic acids containing poly(U) sequences
Iminodiacetate–agarose	Proteins with heavy metal affinity
AMP–agarose	Enzymes with NAD$^+$ cofactors, ATP-dependent kinases

Some binding properties are obvious, such as with the AMP–agarose, but some group-specific resins are not as obvious from their name or structure. Figure 6.4 shows the structure of Cibacron Blue. It is a dye that is linked to agarose-type resins and makes an excellent affinity ligand for enzymes that have nucleotide cofactors, which includes the dehydrogenase family of enzymes.

6.4 Elution of Bound Molecules

Once a protein of interest has bound to the affinity resin, you must find a way to elute it. There are two basic plans for doing this.

High-Salt Elution

Just as we saw with ion-exchange chromatography (IEX), adding a solution of high-salt concentration often disrupts the binding of the ligand and the protein. Although the binding is not based on net charge, almost all binding of enzyme to substrates is based on electrostatic interactions. A very high ionic-strength solution can disrupt those interactions and cause the bound molecule to be released. This can be done in batch style or with a gradient, just as in IEX.

FIGURE 6.4 *Structure of Cibacron Blue 3G*

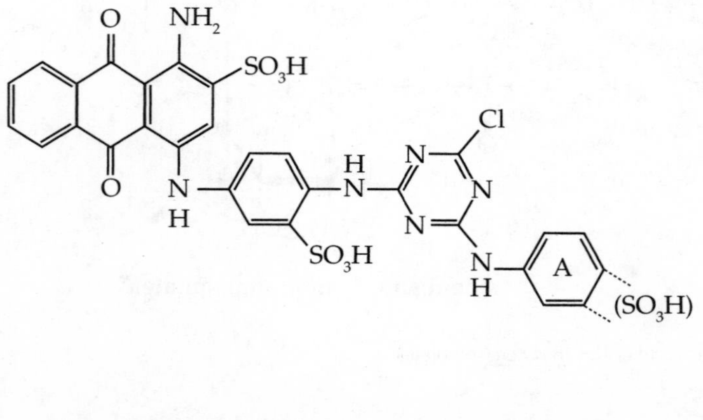

Affinity Elution

Affinity elution is shown in Figure 6.1. Instead of eluting the bound protein with high salt, the bound protein is eluted with its own substrate. If you have LDH bound to an AMP–agarose column, running NADH over the column will likely remove the LDH because the LDH would recognize its own substrate and bind to it rather than the column. The advantage to affinity elution is that it gives another level of specificity to the separation. The disadvantage is that the substrates used for affinity elution are often very expensive.

6.5 Why Is This Important?

Of all the column-purification techniques, affinity chromatography has the greatest potential for increases in purity with a single step. An enzyme for which you can create a very specific ligand will be easy to purify. This approach is currently being used in molecular biology, where genes are being modified so that the protein they express contain a specific tag, such as an N-terminal series of histidines. An affinity column based on chelated nickel ions will bind the histidines very tightly and bind nothing else. Nickel-column manufacturers claim that histidine-tagged proteins can be separated from crude homogenates to complete purity in only one step. This technique is used in Experiment 12 to separate barracuda LDH that is produced by cloning.

Experiment 6

Affinity Chromatography of LDH

In this experiment, you will use affinity chromatography with Cibacron Blue Sepharose to purify LDH. This experiment is a continuation of Experiment 4.

Objectives

Upon successful completion of this lab, you will be able to

- Prepare a Cibacron Blue Sepharose column.

- Assay a sample for LDH, calculate relative activity, and calculate the number of units loaded on the column.

- Wash unbound proteins from the column and monitor with a UV flow-through spectrophotometer.

- Elute LDH, using NaCl, and isolate the active fractions.

- Pool and dialyze elution fractions with significant activity and calculate the yield.

- Explain how Cibacron Blue works to purify LDH.

Experimental Procedures

Preparing Cibacron Blue Sepharose

Materials

Econo columns (1.5 × 15 cm) and accessories

Cibacron Blue Sepharose in 0.02 M sodium phosphate, pH 7.0

Procedure

1. Acquire the columns and accessories that you will need.

2. Acquire the column resin that you will need.

3. De-gas the Cibacron Blue Sepharose with a sidearm flask and vacuum line until bubbles are not coming off.

4. Pour the Cibacron Blue Sepharose to a height of about 10 cm.

5. Stopper the column well and wrap both top and bottom with Parafilm. Store until needed.

Affinity Chromatography of LDH

Materials

Econo columns, 1.5 × 15 cm

Cibacron Blue Sepharose in 0.02 M sodium phosphate, pH 7.0

Fraction collectors

UV monitors

CAPS buffer, 0.15 M, pH 10

NAD^+, 6 mM

Lactate, 0.15 M

Sodium phosphate, 0.02 M, pH 7.0 (wash buffer)

Sodium phosphate, pH 7.0, with 0.2 M, 0.4 M, 0.6 M, 0.8 M, and 1 M NaCl (elution buffers) or just with 1 M NaCl if you will use a gradient

Procedure

1. Acquire your affinity column and dialyzed LDH sample from Experiment 4. If this experiment is being done apart from the LDH-purification series, you will be using another LDH sample for this, and step 2 is unnecessary.

2. If a precipitate has occurred, spin it down with a benchtop centrifuge. Discard the precipitate.

3. Set up fraction collectors, UV monitors, and columns.

4. Assay your LDH fraction to see if any activity has been lost during dialysis and to see how much you are loading on the Cibacron Blue Sepharose column.

5. Load the sample and collect 5-mL fractions. When the sample is loaded, switch to 0.02 M sodium phosphate wash buffer. Continue washing until the UV monitor is back to baseline or you have otherwise determined that all protein not sticking to the column has washed off.

6. Assay the wash fractions to be sure that your LDH bound to the column.

7. Elute with 10-mL aliquots of sodium phosphate plus NaCl, collecting 2-mL fractions, or use a gradient of NaCl from zero to 1 M in 0.02 M NaPhosphate, pH 7.0.

TIP 6.1 Assay the flow-through from the concentrator step to be sure your LDH didn't leak through.

8. Assay the fractions, isolating the peak fractions with significant activity. Again, the UV monitor may be able to help you here.

9. Pool the peak fractions that contain significant activity.

10. Assay your fraction.

Analysis of Results

Experiment 6: **Affinity Chromatography of LDH**

Data

1. Fill in the following table concerning the dialyzed sample that you loaded on the column.

Summary of Dialyzed Sample Loaded on Cibacron Blue Column

Quantity assayed (μL)	
Dilution used (if any)	
$\Delta A / \Delta$ min	
U/mL	
Total milliliters loaded on column	
Total units loaded on column	

2. Fill in the following table for the wash buffer fractions.

Data from Wash Fractions

Fraction No.	Volume Assayed	Δ A/Δ min	U/mL	Total Units

3. Fill in the following table for the elution fractions.

Data from Elution Fractions

Fraction No.	Volume Assayed	$\Delta A/\Delta$ min	U/mL	Total Units

Analysis of Results

1. Fill in the following table concerning your affinity results.

Summary of Affinity Chromatography Results

Fractions pooled	
Volume assayed	
$\Delta A / \Delta$ min	
U/mL	
Total units	
% recovery off of column	
% recovery from beginning of experiment	

2. What percentage of the loaded activity eluted in the wash buffer?

3. What percentage eluted in the elution buffers?

4. Which salt concentration did the most activity elute with? If you used a gradient, estimate the salt concentration from the volume, assuming a linear change from 0 to 1 M.

Experiment 6a

Affinity Chromatography of LDH

In this experiment, you will use affinity chromatography with Cibacron blue Sepharose™ to purify lactate dehydrogenase. This would work well after the ion-exchange chromatography from Experiment 5a.

Objectives

Upon successful completion of this experiment, you will be able to:

- Prepare a Cibacron Blue Sepharose column.

- Assay a sample for LDH, calculate relative activity, and calculate the number of units loaded on the column.

- Wash unbound proteins from the column and monitor with a UV flow-through spectrophotometer.

- Elute LDH using NaCl and isolate the active fractions.

- Pool and dialyze elution fractions with significant activity and calculate yield.

- Explain how Cibacron blue works to purify LDH

Experimental Procedures

Preparing Cibacron Blue Sepharose

Materials

Econo columns (1.5 × 15) and accessories

Cibacron Blue Sepharose in 0.02 M sodium phosphate, pH 7.0

Procedure

1. Acquire the columns and accessories that you will need.

2. Acquire the column resin that you will need.

3. De-gas the Cibacron Blue Sepharose with a sidearm flask and vacuum line until bubbles are not coming off.

4. Pour the Cibacron Blue Sepharose to a height of about 10 cm.

5. Stopper the column well and wrap both top and bottom with Parafilm. Store until needed.

Affinity Chromatography of LDH
Materials

Econo columns (1.5 × 15 cm)

Cibacron Blue Sepharose in 0.02 M sodium phosphate, pH 7.0

Fraction collectors

UV Monitors

CAPS buffer, 0.15 M, pH 10

NAD^+, 6 mM

Lactate, 0.15 M

Sodium phosphate, 0.02 M, pH 7.0 (wash buffer)

Sodium phosphate, pH 7.0 with 0.2 M, 0.4 M, 0.6 M, 0.8 M, and 1 M $NaCl_2$ (elution buffers) or just 1 M NaCl if you will use a gradient.

Millipore or Centricon Spin concentrators.

Procedures

1. Acquire your affinity column and dialyzed LDH sample from Experiment 5a. If this experiment is being done apart from the LDH-purification series, you will be using another LDH sample for this, and step 2 is unnecessary.

2. If a precipitate has occurred, spin it down with a benchtop centrifuge. Discard the precipitate.

3. Set up fraction collectors, UV monitors, and columns as in Experiment 5a. If the tubing is the same size, you should not need to recalibrate the drop count vs. volume. If you are not using a UV monitor or fraction collector, these procedures will have to be modified.

4. Assay your LDH fraction to see if any activity has been lost during dialysis and to see how much you are loading on the Cibacron Blue Sepharose column.

5. Load the sample and collect 5-mL fractions. When the sample is loaded switch to the 0.02 M sodium phosphate wash buffer. Continue washing until the UV monitor is back to baseline or you have otherwise determined that all protein not sticking to the column has washed off.

6. Assay the wash fractions to be sure that your LDH bound to the column.

7. Elute with 10-mL aliquots of sodium phosphate plus NaCl as before, collecting 2-mL fractions, or use a gradient of NaCl from zero to 1 M in 0.02 M NaPhosphate, pH 7.0.

   1 mL

8. Assay the fractions, isolating the peak fractions with significant activity. Again, the UV monitors may be able to help you here.

TIP 6.2 Assay the flow-through from the concentrator step to be sure your LDH didn't leak through.

9. Pool the peak fractions that contain significant activity.

10. If you are going to go on to Experiment 7a, concentrate your pooled fractions down to 1 mL using spin columns according to the manufacturer's instructions.

11. Assay your concentrated fraction. Note this should take a sizeable dilution if everything went according to plan.

12. Store your samples at 4° C.

Analysis of Results

Experiment 6a: **Affinity Chromatography of LDH**

Data

1. Fill in the following table concerning the dialyzed sample that you loaded on the column.

Summary of Dialyzed Sample Loaded on Cibacron Blue Column

Quantity assayed (μL)	
Dilution used (if any)	
$\Delta A / \Delta$ min	
U/mL	
Total milliliters loaded on column	
Total units loaded on column	

2. Fill in the following table for the wash buffer fractions.

Data from Wash Fractions

Fraction No.	Volume Assayed	Δ A/Δ min	U/mL	Total Units

3. Fill in the following table for the elution fractions.

Data from Elution Fractions

Fraction No.	Volume Assayed	Δ A/Δ min	U/mL	Total Units

Analysis of Results

1. Fill in the following table concerning your affinity results.

Summary of Affinity Chromatography Results

Fractions pooled	
Volume assayed	
$\Delta A / \Delta$ min	
U/mL	
Total units	
% recovery off of column	
% recovery from beginning of experiment	

2. What percentage of the loaded activity eluted in the wash buffer?

3. What percentage eluted in the elution buffers?

4. Which salt concentration did the most activity elute with? If you used a gradient, estimate the salt concentration from the volume assuming a linear change from 0 to 1 M.

5. Is there a significant difference between the relative activity of the pooled IEX fractions from Experiment 5a and the dialyzed pellet you used today? If so, how much activity did you lose? How can this be improved?

Additional Problem Set

1. List three enzymes that you expect to bind to Cibacron Blue Sepharose.

2. Compare the advantages and disadvantages of using AMP–Sepharose versus Cibacron Blue Sepharose for the purification of LDH.

3. Bovine LDH has five different isozymes that differ by net charge. How will the charge differences affect its elution from an affinity column? From an ion-exchange column?

4. What are three ways to elute a bound molecule from an affinity column? What are the advantages and disadvantages of each?

5. NADH works well for affinity elution of bound LDH. What are two disadvantages for using it for this purpose?

6. If both NAD^+ and NADH can be used for affinity elution of LDH, which one would you choose and why?

7. You are purifying LDH from a resuspended and dialyzed 65% ammonium sulfate pellet using Cibacron Blue Sepharose chromatography. You load 3 mL of the dialyzed pellet on the column and wash with 50 mL of 0.02 M sodium phosphate, pH 6.0, collecting 5-mL fractions. You then switch to an elution buffer containing 1 mM NADH and collect 3-mL fractions. You assay for LDH via the standard assay used in Experiment 6 and see the following results:

Data for Problem 7 LDH Purification

Fraction Assayed	Volume of Sample Assayed (μL)	Dilution Used	Δ A/Δ min
Dialyzed 65%	50	100/1	0.508
Wash fraction 1	100	—	0.03
Wash fraction 2	100	—	0.08
Wash fraction 3	100	—	0.10
Wash fraction 4	100	—	0.07
Wash fraction 5	100	—	0.02
Wash fraction 6	100	—	0
Elution fraction 1	100	—	0
Elution fraction 2	100	—	0.15
Elution fraction 3	50	—	0.40
Elution fraction 4	10	—	0.50
Elution fraction 5	20	10/1	0.05
Elution fraction 6	20	10/1	0.16
Elution fraction 7	10	10/1	0.12
Elution fraction 8	20	10/1	0.10

(continued)

Fraction Assayed	Volume of Sample Assayed (μL)	Dilution Used	$\Delta A/\Delta$ min
Elution fraction 9	10	—	0.24
Elution fraction 10	50	—	0.15
Elution fraction 11	100	—	0.05
Elution fraction 12	100	—	0

a. Calculate the number of units and units/milliliter for each fraction.

b. What percentage of the LDH loaded on the column was recovered?

c. Which fractions would you pool to continue on to another purification step? Justify your answer.

d. What could you do to increase the enzyme yield from this column?

Webconnections

For a list of websites related to the material covered in this chapter, go to **Webconnections** at the *Experiments in Biochemistry* site on the Brooks/Cole Publishing website. You can access this page at http://www.brookscole.com and follow the links from the chemistry page.

References and Further Reading

Bannikova, G. E., V. P. Varlamov, M. L. Miroshnichenko, and E. A. Bonch-Osmolovskaya. "Isolation of Thermostable Phosphatase from the Hyperthermophilic Archaeon *Thermococcus pacificus* by Immobilized Metal Affinity Chromatography." *Biochemistry and Molecular Biology International* 44, no. 2 (1998).

Bevilacqua, P. C., C. X. George, C. E. Samuel, and T. R. Cech. "Binding of the Protein Kinase PKR to RNAs with Secondary Structure Defects." *Biochemistry* 37, no. 18 (1998).

Blanchard, S. C., D. Fourmy, R. G. Eason, and J. D. Puglisi. "rRNA Chemical Groups Required for Aminoglycoside Binding." *Biochemistry* 37, no. 21 (1998).

Boyer, R. F. *Modern Experimental Biochemistry*. Menlo Park, CA: Addison-Wesley, 1993.

Campbell, M. *Biochemistry*. Philadelphia: Saunders, 1998.

Canduri, F., R. J. Ward, W. F. de Azevedo, Jr., R. A. Gomes, and R. K. Arni. "Purification and Partial Characterization of Cathepsin D from Porcine Liver Using Affinity Chromatography." *Biochemistry and Molecular Biology International* 45, no. 4 (1998).

Dharmawardana K. R., and P. E. Bock. "Demonstration of Exosite I-Dependent Interactions of Thrombin with Human Factor V and Factor Va Involving the

Factor Va Heavy Chain; Analysis by Affinity Chromatography Employing a Novel Method for Active-Site-Selective Immobilization of Serine Proteases." *Biochemistry* 37, no. 38 (1998).

Dryer R. L., and G. F. Lata. *Experimental Biochemistry*. New York: Oxford University Press, 1989.

Eisen, C., C. Meyer, R. Dressendorfer, C. Strasburger, H. Decker, and M. Wehling. "Biotin-Labelled and Photoactivatable Aldosterone and Progesterone Derivatives as Ligands for Affinity Chromatography, Fluorescence Immunoassays and Photoaffinity Labelling." *European Journal of Biochemistry* 237, no. 2 (1996).

Glerum D. M., and A. Tzagoloff. "Affinity Purification of Yeast Cytochrome Oxidase with Biotinylated Subunits 4, 5, or 6." *Analytical Biochemistry* 260, no. 1 (1998).

Jack, R. C. *Basic Biochemical Laboratory Procedures and Computing*. New York: Oxford University Press, 1995.

Kline, T. *Handbook of Affinity Chromatography*. New York: M. Dekker, 1993.

Labrou, N. E., and Y. D. Clonis. "Simultaneous Purification of L-Malate Dehydrogenase and L-Lactate Dehydrogenase from Bovine Heart by Biomimetic-Dye Affinity Chromatography." *Bioprocess Engineering* 16, no. 3 (1997).

Maier, T., N. Drapal, M. Thanbichler, and A. Bock. "Strep-Tag II Affinity Purification." *Analytical Biochemistry* 259, no. 1 (1998).

Mortensen, U. H., H. R. Stennick, and K. Breddam "Reversed-Flow Affinity Elution Applied to the Purification of Carboxypeptidase Y." *Analytical Biochemistry* 258, no. 2 (1998).

Muller, K. M., K. M. Arndt, K. Bauer, and A. Pluckthun. "Tandem Immobilized Metal-Ion Affinity Chromatography/Immunoaffinity Purification of His-Tagged Proteins." *Analytical Biochemistry* 259, no. 1 (1998).

Nagata, Y., K. Maeda, and R. K. Scopes. "NADP Linked Alcohol Dehydrogenases from Extreme Thermophiles: Simple Affinity Purification Schemes and Comparitive Properties of the Enzymes from Different Strains." *Bioseparation* 2 (1992).

O'Shannessy, K., J. Scoble, and R. K. Scopes. "A Simple and Economical Procedure for Purification of Muscle Lactate Dehydrogenase by Batch Dye-Ligand Adsorption." *Bioseparation* 6, no. 2 (1996).

Robyt, J. F., and B. J. White. *Biochemical Techniques*. Long Grove, IL: Waveland Press, 1990.

Scopes, R. K. "Affinity Elution Chromatographic Procedures." *Journal of Biochemistry* 161 (1977).

Wittlin, S., J. Rosel, and D. R. Stover. "One-Step Purification of Cathepsin D by Affinity Chromatography Using Immobilized Propeptide Sequences." *European Journal of Biochemistry* 252, no. 3 (1998).

Chapter 7
Gel Filtration Chromatography

Introduction

In this chapter, we address the third and final type of column chromatography that we will use. **Gel filtration,** *also called molecular sieving and size exclusion chromatography, is a preparative and analytical technique that allows us to purify macromolecules away from others of different sizes. Information about the molecular weight of a protein can also be determined.*

7.1 Gel Filtration

Proteins and other macromolecules can be separated based on molecular weight (MW) by using a cross-linked porous gel. Gels have various degrees of cross-linking that allow certain sized molecules to pass through while others cannot. Thus, some gels are good at separating large molecules, and others are better for smaller ones. The degree of retardation of a particle is related to the molecular weight and shape. Some molecules will be excluded from the gel, and others will be included. Those that are excluded will pass through the column *faster* because the distance they have to travel is reduced. The smaller molecules will take a more convoluted path through the column and elute later. Figure 7.1 demonstrates this process.

This technique is useful for many purposes. It is nondestructive, so you do not lose any of your sample. It can be used to determine the molecular weight of a compound in its native conformation. It can be used to purify a compound from other compounds of differing sizes.

7.2 Types of Supports

There are three common types of supports for gel filtration: dextran, polyacrylamide, and agarose. Dextran is a polysaccharide-based resin with glucose residues linked $\alpha\ 1 \rightarrow 6$ with branches of $\alpha\ 1 \rightarrow 3$. These are normally sold under the trade name of Sephadex by Pharmacia or Sigma Chemicals. If you have a bottle of Sephadex, it will be labeled with something like G-100.

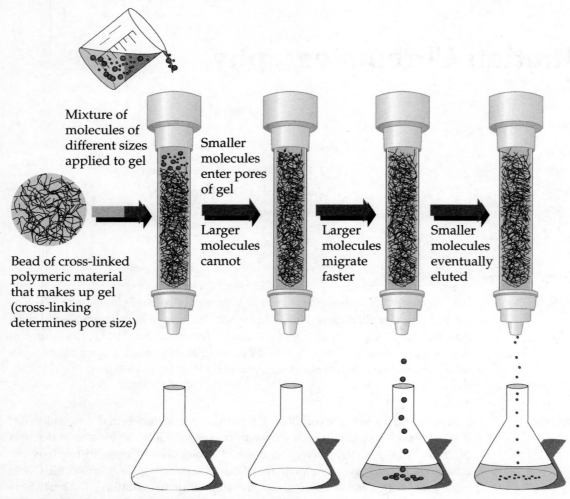

FIGURE 7.1 *Basics of gel filtration chromatography*

The G refers to the amount of water gained by the dehydrated gel when it is allowed to swell. Sephadex comes in many different sizes, all of which have different levels of cross-linking (that is, pore size) and therefore separate molecules in a different range of molecular weights. Table 7.1 gives some properties of the Sephadex gels.

Sephadex gels have different **exclusion limits,** which is the molecular weight limit for entrance into the gel beads. If a gel has an exclusion limit of 80,000, then all proteins of MW 80,000 or bigger will not have access to the beads. These can be seen from the higher number on the fractionation range in Table 7.1. (Sephadex G-150 and G-200 have been discontinued.)

Polyacrylamide gels are made of beads composed of the same material used to make acrylamide gels for electrophoresis. These are usually produced under the trade name Bio Gel, by Bio Rad Laboratories, or Sephacryl, by Sigma.

TABLE 7.1 *Properties of Sephadex Gels*

Dextran Type	Fractionation Range (daltons)	Water Regain (mL/g dry gel)	Bed Volume (mL/g dry gel)
G-10	0–700	1.0	2–3
G-15	0–1500	1.5	2.5–3.5
G-25	1,000–5,000	2.5	4–6
G-50	1,500–30,000	5.0	9–11
G-75	3,000–80,000	7.5	12–15
G-100	4,000–150,000	10	15–20
G-150	5,000–300,000	15	20–30
G-200	5,000–600,000	20	30–40

Agarose beads are made of the same material used for agarose gel electrophoresis. These are usually sold under the trade name Sepharose, also by Pharmacia and Sigma.

7.3 Determining Molecular Weight

When native, globular proteins are run on the appropriate size of gel filtration resin, they separate according to their molecular weights. If we plot the log of the molecular weight versus the elution volume from the column, we should see a straight line. The **elution volume,** V_e, is the volume of liquid that is collected from the moment the sample is applied to the column until the sample is collected from the column. Because each column is different in length, width, particle size, level of hydration, and so on, comparing elution volumes for the same molecule on different columns is difficult, even if these columns theoretically had the same material. There are several ways of overcoming this difficulty, all of which are similar to taking an Rf, as we saw with thin-layer chromatography. Figure 7.2 shows a typical calibration curve for a gel, using the ratio of the elution volume to the void volume of the column (see Section 7.4). Note that each column should be individually calibrated for the greatest accuracy.

7.4 Distribution Coefficients

An equation can be set up for molecules moving through a column:

$$K_d = \frac{(V_e - V_o)}{(V_i)}$$

where V_e = elution volume for the solute of interest, in this case a protein whose molecular weight that we want to determine.

V_o = void volume, which is the elution volume of a compound that is completely excluded from the gel. This is usually

FIGURE 7.2 *Calibration curve for a Sephadex gel*

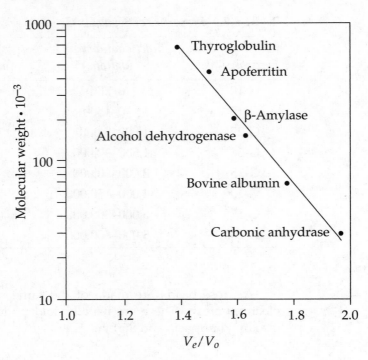

determined by measuring the volume it takes a sample of blue dextran to elute.

V_i = inner volume, the volume of liquid inside the gel bead.

Because it is difficult to measure V_i accurately, the value for the K_d (the real distribution coefficient) is also difficult to measure accurately. A simplifying equation using the total column volume, V_t, is usually used instead.

V_t = total column volume, which equals $V_i + V_o + V_g$, where V_g is the volume of the gel bead itself, not including any solvent. If we assume that the volume of the gel is insignificant compared to the column volume, we can say that $V_t = V_i + V_o$, or by rearranging, $V_i = V_t - V_o$.

V_t can be calculated by taking the actual volume of the column as $\pi r^2 h$ or by measuring the amount of water necessary to fill the column to the point where the gel was packed. Assuming that the volume of the gel is negligible will also allow us to calculate V_t another way. If V_t is considered to be $V_o + V_i$, then it is also the volume available to a molecule that has access to all pores in the gel. A small molecule, such as DNP–aspartate is used to measure this.

A working coefficient, called K_{avg} is often used:

$$K_{avg} = \frac{V_e - V_o}{V_t - V_o}$$

This distribution coefficient means that a solute is distributed between the two parts of the column: the space within and the space between the porous beads. Graphing log MW versus K_d or K_{avg} gives a straight line for those proteins falling within the linear separation limits of the gel. Because the difference between K_d and K_{avg} is small and because some sources define K_d the same way we define K_{avg}, we will use K_d throughout to mean K_{avg} as defined earlier.

The K_d's for many proteins have been established for the common types of gel filtration media. To find the molecular weight of an unknown protein, its K_d can be determined, and that can be compared to known K_d's of proteins on the same gel matrix by using a graph. Although not as accurate as establishing your own calibration curve for your column, this technique allows you to estimate the molecular weight of an unknown without running multiple standards.

Because the shape is not taken into consideration, the molecular weight is an estimate. Two proteins that weigh exactly 60,000 may elute differently on the gel if their shapes are quite different. For example, a cigar-shaped protein has an effective radius much larger than a spherical one of the same weight.

To calculate the molecular weight of an unknown protein by this method, calculate the V_o, V_e, and V_t of the column. To do this, run a large molecule, such as blue dextran, to measure V_o. Next apply the blue dextran to the column and see what volume it takes for it to come off the column. Then do the same thing with the protein whose K_d that you are trying to measure. This is the V_e. To calculate V_t, measure the volume of water it takes to fill up the column to the height of the packed gel. In Experiment 7, we will calculate V_t by using a tiny molecule, a DNP-linked amino acid, to measure V_t. After running all three compounds through the column, make a graph such as the one shown in Figure 7.3. The volumes of the three peaks indicate the void volume, elution volume, and total volumes, respectively. Then calculate K_d according to the equation given previously.

PRACTICE SESSION 7.1   Using the information in Figure 7.3, what is the K_d for the protein of interest?

The void volume is the volume at which blue dextran elutes, which is 20 mL. The elution volume is that of the protein, 60 mL, and the total volume is that of the DNP–amino acid, 120 mL. Note that we took the fraction

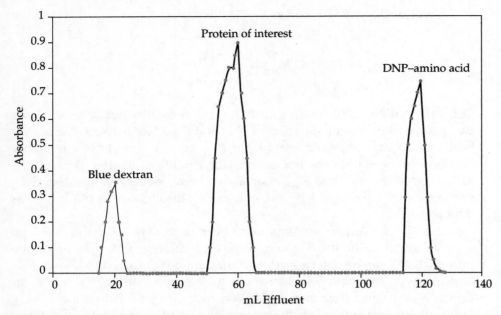

FIGURE 7.3 *Absorbance versus volume for three components separated by gel filtration*

off the column that had the highest absorbance for those compounds to represent the peaks. Thus, the K_d is calculated as follows:

$$K_d = \frac{60 - 20}{120 - 20} = 0.4$$

PRACTICE SESSION 7.2   What is the molecular weight of the protein of interest?

To answer this question, you must be able to graph log MW of known proteins versus their K_d values. Assuming that you were provided with a table of K_d values and molecular weights, you can make a graph such as the one shown in Figure 7.4. This is a semilog graph of molecular weight versus K_d. Then interpolate the K_d of your protein of interest off the graph. There is a known protein with a K_d of 0.4; its molecular weight is 20,000.

ESSENTIAL INFORMATION

Gel filtration is a powerful technique for purification and analysis of biomolecules. The most common gels are those produced by Pharmacia and Sigma, which are called Sephadex, Sepharose, and Sephacryl. Each gel has a pore size that optimizes it for certain sizes of molecules. All molecules too big elute at the same time in the void volume. All molecules too small elute together in the total volume. The molecules that come out in-between are in the fractionation range of the gel. These separate linearly according to the log of their molecular weights.

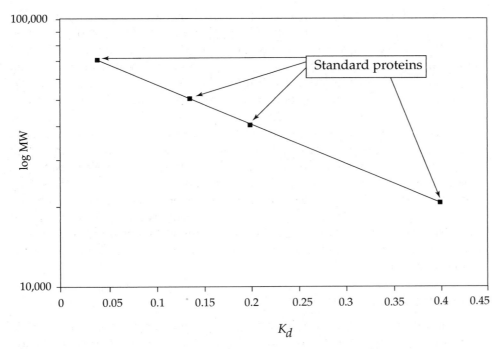

FIGURE 7.4 *Log MW versus K_d for proteins separated by gel filtration*

7.5 Why Is This Important?

Gel filtration is a very common, quick, and easy technique that is used in almost every subfield of biochemistry. It is used for purifying proteins, nucleic acids, and carbohydrates—to name just a few. It is also used to remove salt or other small molecules as a prelude to continuing purification. For example, if you use ammonium sulfate to salt out a protein, the sample you get contains a lot of salt. Putting the sample through the next step may be difficult or impossible until the salt is removed. In the experiment that follows, you will see how a large molecule (blue dextran) and a small molecule (DNP–aspartate) separate very quickly on this column. The same is true for a protein and excess salt.

7.6 Expanding the Topic

Calculating Elution Volumes

Many students find it difficult to calculate elution volumes for a couple of reasons. First, you often start a chromatography experiment by collecting the eluant into a graduated cylinder instead of test tubes. You do this to avoid wasting time collecting fractions when you know that samples are not coming out yet. No sample can precede the void volume, so if you know the void volume is 30 mL for a column, there is no reason to collect fractions until close to that value. This means that you will have one volume collected in the graduated cylinder and more collected in the fractions. To calculate any elution volume, you must include all of the volume collected (the graduated cylinder and the fractions).

Second, it is often unclear what volume to record for the eluting samples. Samples tend to spread out with any type of chromatography, so even if only 0.5 mL of a protein solution was applied, the protein will elute from the column in several milliliters. What volume will you call the elution volume? The V_o is defined as the volume of the liquid outside the beads. If you load blue dextran on a Sephadex column to measure V_o, the actual volume of the column, which excludes the gel volume and the interior bead volume, is the volume at which the blue dextran first starts to come off. Some researchers prefer to plot absorbance versus elution volume as in Figure 7.3; then they extrapolate the leading edge of the blue dextran peak back to where it crosses the x axis and use that volume as the V_o. However, an accepted practice is to use the peak absorbance fraction to calculate the volume. This is easier because it relies on positive data for maximum absorbencies instead of trying to calculate where the first molecule of blue dextran came out. The elution volumes for the other components are calculated the same way.

Column Calibration

There are different ways to determine the molecular weight of a molecule through gel filtration. If you are going to use a column repeatedly, the most accurate way is to take proteins of known molecular weight and run them as standards. This gives a unique standard curve of log MW versus V_e that can be used for that column and is definitely the safest way to go. If you wish to compare your data to those obtained from another experimenter, then you will need to have some type of relative elution volume. We have seen two ways of doing that in this chapter. One is to plot V_e/V_o. The other is to calculate a K_d, or K_{avg}. To determine the latter, you need to know the total column volume, V_t, which can be determined in several ways too. The easiest is to measure the liquid volume contained by the column at the height of the packed gel. You can also use an algebraic measurement based on the height and internal diameter. The elution volume of a molecule that has a K_d of 1 will also be nearly identical to V_t.

Experiment 7

Gel Filtration Chromatography

In this experiment, you will separate a three-component mixture using gel filtration chromatography on Sephadex G-75. One component is a protein, and you will determine its molecular weight. This experiment can be done with or without fraction collectors and UV monitors, depending on your available equipment.

Prelab Questions

1. What is the purpose of the slow-loading procedure from steps 1–4? Why do we put small quantities of buffer on top of the column?

2. Why must the column never run dry?

3. What must be done to the column fractions before they are assayed?

4. How will you know when it is time to stop running buffer through the column at the end of the experiment?

Objectives

After successful completion of this lab, you will be able to

- Properly load a sample on the Sephadex gel without disturbing the gel bed.

- Measure the peak elution volumes of blue dextran, DNP–aspartate, and a given protein.

- Determine the K_d of an unknown protein.

- Calculate the native molecular weight given the K_d you calculated and a table of known K_d's of standard proteins.

- Use semilog paper to plot molecular weights.

Experimental Procedures

Materials

Sephadex G-75 (regular mesh), hydrated in 0.05 M TRIS-HCl, pH 7.5

Elution buffer, 0.05 M TRIS-HCl, pH 7.5

Separation mixture, 6 mg blue dextran, 0.6 mg DNP–aspartate, 12 mg mystery protein, in 1 mL of elution buffer

Chromatographic column, 1.5 cm × 60 cm

Column peripherals, if any, such as fraction collectors and pumps

Methods

1. Drain the buffer that is on top of the column until the meniscus is just starting to enter the resin. If there is too much buffer on top, pull off some of it with a Pasteur pipet before draining the rest. Place a graduated cylinder under the column to catch the effluent.

2. Load 0.5 mL of the separation mixture onto the column. This part must be done *extremely carefully!* Your results will be wrong if you rush the loading process. The best way is to use a Pasteur pipet with a pipet pump. Slowly circle the pipet tip around the inside of the column near the resin to avoid making a pit in one spot of the column bed.

3. Let the mixture run into the column while collecting the effluent into a graduated cylinder.

4. Do not add any more buffer to the top of the column until the mixture has completely entered. Then add small quantities and let the column run some more.

5. When the mixture is safely in the gel (2 cm from the top of the gel bed), fill the column to the top with elution buffer. *Never let the top of the gel bed get completely dry.*

6. When the blue dextran is getting close to exiting the column, remove the graduated cylinder and record the volume collected to this point. Begin to collect 1-mL fractions into small test tubes.

Analyzing Column Data without UV Monitors

1. To establish the peak of elution for the blue dextran, add 1.5 mL of water to the fractions and assay with a spectrophotometer set at 650 nm. This

---

TIP 7.1   Beware of the loading process! Slow loading with a Pasteur pipet is the only way to do a decent job. Be very careful that you do not drop the pipet into the resin or that the pipet does not fall out of the pipet pump while you are loading.

is the wavelength of maximum absorbance for blue dextran. Find the fraction with the highest absorbance.

2. Continue the elution and find the peak of elution of the unknown protein by measuring the absorbance of the tubes at 500 nm (after adding 1.5 mL of water). While all proteins have a wavelength of maximum absorbance at 280 nm, this one happens to be reddish brown and can be detected with visible spectrophotometry at 500 nm.

3. Continue eluting the column and collecting the fractions until the DNP–aspartate has completely eluted and the column is clean. Find the peak elution fraction by measuring the absorbance at 440 nm (after adding 1.5 mL of water).

Analyzing Column Data with a Flow-Through UV Monitor

1. Although the three components all have different visible wavelengths of maximum absorbance, they can all be detected at 280 nm. Use the UV monitor and chart recorder to create a graph of absorbance at 280 nm versus elution volume.

2. Determine the volumes that correspond to the three peaks.

Name _____   Section _____

Lab partner(s) _____   Date _____

Analysis of Results

Experiment 7: **Gel Filtration Chromatography**

Data

Turn in all of your raw data concerning the cumulative volume of buffer eluted from the column and the absorbance you recorded.

Calculations

1. On a single graph, plot the absorbance of the fractions versus the volume taken at that fraction. For example, if you collected 20 mL into the graduated cylinder and then began collecting 1-mL fractions and your first dextran blue fraction that had an absorbance on the fourth fraction, then plot the absorbance versus 20 mL + 4 mL = 24 mL.

2. What were the elution volumes that gave the highest absorbencies for the blue dextran, the unknown protein, and the DNP–aspartate?

   Absorbance of blue dextran _____ cumulative volume _____ = V_o

   Absorbance of mystery protein ___ cumulative volume _____ = V_e

   Absorbance of DNP–aspartate ___ cumulative volume _____ = V_t

3. Calculate the K_d for the unknown protein.

$$K_d = \frac{V_e - V_o}{V_t - V_o}$$

4. Use the information in Table 7.2 to make a graph of $\log_{10}$ MW versus K_d. (Now might be a good time to review Section 1.8 on graphing.)

TABLE 7.2 K_d's for Some Known Proteins on Sephadex G-75

Protein	Molecular Weight	K_d
Trypsin inhibitor (pancreas)	6,500	0.70
Trypsin Inhibitor (lima bean)	9,000	0.60
Cytochrome c	12,400	0.50
α-lactalbumin	15,500	0.43
α-chymotrypsin	22,500	0.32
Carbonic anhydrase	30,000	0.23
Ovalbumin	45,000	0.12

5. Interpolate from the graph and determine the molecular weight of the unknown protein. You do *not* need to determine its identity.

molecular weight of mystery protein =

Questions

1. Why do large proteins come out of a gel filtration column faster than small ones? Doesn't it make more sense for small things to move more quickly through a gel?

2. Can you use Sephadex G-75 to separate alcohol dehydrogenase (MW 150,000) from β-amylase (MW 200,000)? Why or why not?

3. Can you use Sephadex G-75 to separate alcohol dehydrogenase from bovine serum albumin? Why or why not?

4. How can you use the equation for K_d and the data in Table 7.2 to calculate the exclusion limit for Sephadex G-75?

5. What characteristics of a column of Sephadex G-75 determine if you can effectively separate cytochrome c from α-lactalbumin?

Experiment 7a

Gel Filtration Chromatography of LDH

In this experiment, you will do the final column purification of LDH, using gel filtration chromatography with Sephadex G-150 or Sephacryl 200.

Prelab Questions

1. What is the purpose of the slow-loading procedure? Why do we put small quantities of buffer on top of the column?

2. Where do you expect LDH to elute with Sephadex G-150 or Sephacryl 200? Why?

3. Why must the column never run dry?

Objectives

After successful completion of this lab, you will be able to

• Properly load a sample on the Sephadex gel without disturbing the gel bed.

• Properly run a Sephadex gel without letting it run dry.

• Locate and pool fractions containing LDH, using a UV monitor and enzyme assay.

• Estimate the native molecular weight of LDH, given a previously constructed graph of log MW versus V_e/V_o.

Experimental Procedures

Preparing Gel Filtration Columns

Materials

Econo columns, 1.5 × 60 cm

Sephadex G-150 (regular mesh) or Sephacryl 200 previously swelled in 0.05 M sodium phosphate, pH 7.5

198   **Chapter 7** / *Gel Filtration Chromatography*

Procedure

1. Get the columns and accessories that you will need.

2. Get the column resin that you will need.

3. De-gas the resin with a sidearm flask attached to a vacuum line until bubbles are not coming off.

4. Remove fine particles of resin in a graduated cylinder by siphoning off the last 5–10% of the volume of fuzzy particles.

5. Carefully pour the column with the resin. Avoid having the column completely pack before you have to add more. You want to get a smooth application of resin to the system until it is tightly packed to near the top of the column.

6. Stopper the column well and wrap both top and bottom with Parafilm. Store until needed.

7. Check the column several hours before you need to use it in case bubbles have formed or it has leaked and run dry.

Gel Filtration Chromatography of LDH

Materials

Econo columns, 1.5 × 60 cm, packed with Sephadex G-150 or Sephacryl 200

Fraction collectors

UV monitors

0.05 M sodium phosphate, pH 7.5 (elution buffer)

Your purified LDH

Procedure

1. Get your Sephadex column and your concentrated LDH sample.

2. Assay your sample to ensure that it has not lost activity during the storage process.

3. Assemble the column and peripheral hardware, setting the volume of the fraction collector to 1 mL.

4. Drain off the excess buffer that is on top of the gel, being very careful not to disturb the gel bed or let the resin run dry. This column is the most sensitive to the loading and running technique that we have used.

5. Carefully load your concentrated LDH on the gel. From this point on, every drop counts as elution volume.

6. When the LDH sample is just entering the resin, slowly add 1 mL of elution buffer and let that drain in. Add another 1 mL and repeat this process until 5 mL have been added 1 mL at a time. Now *slowly* fill up the reservoir and keep the column running.

7. Assay your fractions and locate the ones containing enzyme, pooling those fractions that have significant activity. Assay the pooled fractions before proceeding to step 8.

8. Because this is the last purification step, you must perform Experiment 3b, the protein assays, before completing the final purification table.

Analysis of Results

Experiment 7a: **Gel Filtration Chromatography of LDH**

Data

1. Fill in the following table concerning the concentrated sample you loaded on the column.

Summary for Concentrated/Dialyzed Sample

Quantity assayed (µL)	
Dilution used (if any)	
$\Delta A / \Delta$ min	
U/mL	
Total milliliters loaded on column	
Total units loaded on column	

2. Turn in all raw data concerning the cumulative volume of buffer eluted from the column and the UV absorbance that you recorded.

3. Record a void volume for the column. This may have been given to you, or you may have been asked to do the void volume yourself.

4. Fill in the following table for the elution fractions.

Data for Elution Fractions

Fraction No.	Volume Assayed	Δ A/Δ min	U/mL	Total Units

Analysis of Results

Fill in the following table concerning your gel filtration results.

Summary of Gel Filtration Results

Fractions pooled	
Volume assayed	
$\Delta A / \Delta$ min	
U/mL	
Total units	
% Recovery off of column	
% Recovery from beginning of experiment	

Calculations

1. Calculate the V_e / V_o ratio for LDH.

2. Using the graphs shown in Figures 7.5 or 7.6, make a rough estimate of the LDH molecular weight. Is your answer reasonable?

FIGURE 7.5 *Log MW versus V_e/V_o for Sephadex G-150*

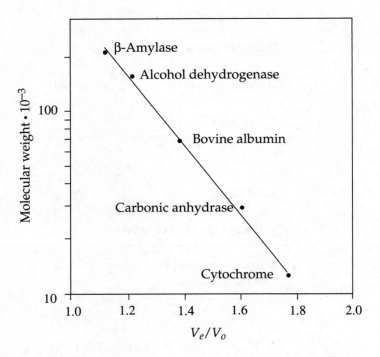

FIGURE 7.6 *V_e/V_o versus log MW for Sephacryl resins*

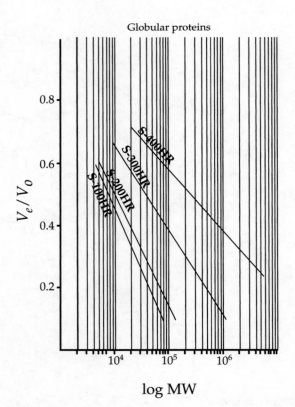

3. Using the following table, finish your enzyme purification table.

Final Purification Table

Fraction	U/mL	Total Units	%Recovery	Protein (mg/mL)	Specific Activity	Fold Purification
Crude						
20,000 × g, supernatant						
65% AS pellet						
Dialyzed 65% AS pellet						
Pooled IEX fractions						
Dialyzed IEX fractions						
Pooled Cibacron Blue fractions						
Concentrated Cibacron Blue fractions						
Pooled Sephadex fractions						
Concentrated Sephadex fractions						

Questions

1. What is the final percent recovery of LDH that you saw? Is this reasonable?

2. What is the final fold purification that you saw? Is this reasonable?

3. Is there a particular place in the experiment where you lost too much activity?

4. What is the most effective step in the LDH purification?

5. What is the least effective step in the LDH purification?

6. If there are still contaminating proteins with your LDH, what are their physical characteristics?

7. Give one example of a protein that might still be with your LDH.

8. Why do large proteins come out of a gel filtration column faster than small ones? Doesn't it make more sense for small things to move more quickly through a gel?

9. Can you use Sephadex G-75 to separate alcohol dehydrogenase (MW 150,000) from β-amylase (MW 200,000)? Why or why not?

10. Can you use Sephadex G-75 to separate alcohol dehydrogenase from bovine serum albumin? Why or why not?

11. How can you use the data shown in Figure 7.5 to calculate the exclusion limit for Sephadex G-150?

12. Does the number 150 in Sephadex G-150 mean that it has an exclusion limit of 150 kD?

Additional Problem Set

1. If you chromatograph an oblong protein, why does it appear to be larger than a spherical protein of equal mass?

2. Using Table 7.1, explain how to prepare an 80-mL column of an appropriate Sephadex gel needed to separate proteins ranging in size from bovine serum albumin to β-amylase.

3. Explain two advantages of using K_d values instead of calculating the raw elution volume of your protein on your gel.

4. Explain how to determine the exclusion limit of a gel of unknown origin.

5. What is the possible range of a K_d value? How does the value relate to the size of the protein?

6. Why is sodium azide often added to the storage buffer for a gel filtration column?

7. Why are gel filtration columns usually long and narrow whereas ion-exchange columns are short and fat?

Webconnections

For a list of web sites related to the material covered in this chapter, go to **Webconnections** at the *Experiments in Biochemistry* site on the Brooks/Cole Publishing web site. You can access this page at http://www.brookscole.com and follow the links from the chemistry page.

References and Further Reading

Ahern, H. "Chromatography." *The Scientist* 10, no. 5 (1996).

Boyer, R. F. *Modern Experimental Biochemistry*. Menlo Park, CA: Addison-Wesley, 1993.

Campbell, M. K. *Biochemistry*. Philadelphia: Saunders, 1998.

Dryer, R. L., and G. F. Lata, *Experimental Biochemistry*. New York: Oxford University Press, 1989.

Robyt, J. F., and B. J. White. *Biochemical Techniques*. Long Grove, IL: Waveland, Press, 1990.

Sigma Chemical Company. *Biochemicals, Organic Compounds for Research, and Diagnostic Reagents*, 1995.

Stenesh, J. *Experimental Biochemistry*. Boston: Allyn and Bacon, 1984.

Tsao, G. T. *Chromatography*. San Diego: Elsevier Science, 1991.

Whitaker, J. R. "Determination of Molecular Weights of Proteins by Gel Filtration on Sephadex™." *Analytical Chemistry* 35, no. 12 (1963).

Chapter 8

Enzyme Kinetics

TOPICS

Introduction

In this chapter, we explore the nature of enzyme kinetics, which is the study of enzyme rates and their dependence on concentrations of enzyme, substrate, and the kinetic rate constants.

8.1 Reaction Rates

Enzyme kinetics is the study of enzyme rates and how these rates are affected by enzyme concentration substrates, and any inhibitors or activators. The reaction rate is expressed as a change in the concentration of a reactant or product during a given time interval.

If a reaction can be written as

$$A + B \rightarrow P$$

then the rate can be expressed in terms of either the appearance of the product P or the disappearance of either of the reactants A or B. Thus,

$$\text{rate} = \frac{\Delta [P]}{\Delta t} = \frac{-\Delta [A]}{\Delta t} = \frac{-\Delta [B]}{\Delta t}$$

It has been established that the reaction rate at a given time is proportional to the product of the concentration of the reactants raised to an appropriate power,

$$\text{rate} \, \alpha [A]^a[B]^b \quad \text{or} \quad \text{rate} = k[A]^a[B]^b$$

where k is a proportionality constant called a rate constant and a and b are constants that must be determined experimentally.

8.2 Order of Reactions

The exponents in the rate equation are usually small whole numbers such as 0, 1, or 2. The values are dependent on the number of molecules of reactants and products involved. The values can often be deduced from the balanced chemical equation but may also be more complicated to determine.

If a reaction with the equation

$$A \rightarrow B$$

can be shown to have the rate equation

$$\text{rate} = k[A]^1$$

then it is said to be **first order** with respect to A. The rate at any time is governed by the rate constant times the concentration of A present at that time. A good example of this type of reaction is radioactive decay. The rate of radioactive decay is always based on the amount of radioactive substance present times a decay constant.

If the rate of a reaction

$$A + B \rightarrow C + D$$

is governed by the rate equation

$$\text{rate} = k[A]^1[B]^1$$

then the reaction is *first order* with respect to A, *first order* with respect to B, but **second order** overall. This rate would go faster if the concentrations of either A or B or both are increased.

8.3 The Michaelis–Menten Approach

The most often seen model for enzyme kinetics was proposed in 1913 by Michaelis and Menten. Although it has undergone much revision, it is still the basic one used for all nonallosteric enzymes.

An enzyme-catalyzed reaction can be written

$$E + S \underset{k_2}{\overset{k_1}{\rightleftharpoons}} ES \underset{k_4}{\overset{k_3}{\rightleftharpoons}} E + P$$

where   E   =   free enzyme

S   =   free substrate

ES   =   enzyme–substrate complex

P   =   product

k_n =   individual rate constants

Usually we perform these experiments for short periods so that product P does not significantly build up. Therefore, there is no back reaction, k_4. The equation then simplifies to

$$\mathbf{E} + \mathbf{S} \underset{k_2}{\overset{k_1}{\rightleftharpoons}} \mathbf{ES} \xrightarrow{k_{cat}} \mathbf{E} + \mathbf{P}$$

where k_{cat} is the catalytic rate constant for breakdown to product.

The rate of an enzyme-catalyzed reaction is dependent on the concentration of enzyme and of substrate, but the relationships are not the same for both. If we plot rate v versus [E], we get a line as shown in Figure 8.1. This has important implications for our ability to do an enzyme kinetic experiment. Any change in the amount of enzyme will change the rate that we observe, so the enzyme must be pipetted very carefully.

However, if we make a similar plot of v versus [S], we get a different graph (Figure 8.2). The rate is dependent on the amount of substrate present, but the line is not straight. At low substrate concentrations, the line is straight, and the rate is first order with respect to S. Then it flattens out and eventually reaches zero order with respect to S. This means that at high [S], the rate is independent of the amount of S present. This is a hyperbolic curve, with an asymptote at what we call V_{max}. V_{max} is the

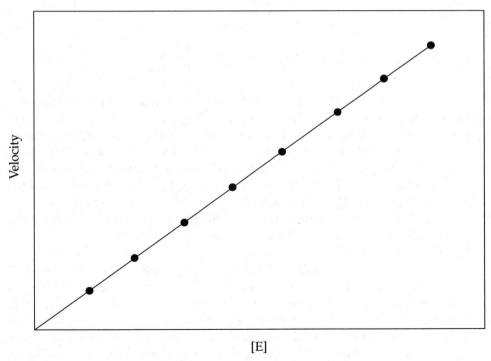

FIGURE 8.1 *Reaction velocity versus enzyme concentration*

FIGURE 8.2 *Velocity versus substrate for the Michaelis–Menten model*

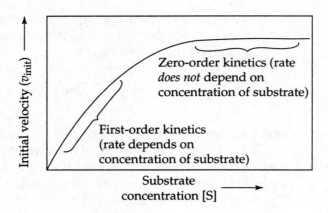

theoretical maximum velocity, which occurs at infinitely high [S]. *At high [S], the velocity increases less with each increase of S.*

To account for this observation of kinetic properties, the Michaelis–Menten equation was derived:

$$v = \frac{V_{max}[S]}{K_m + [S]}$$

where v = initial velocity

V_{max} = maximum velocity

[S] = substrate concentration

K_m = Michaelis constant $= \dfrac{k_2 + k_{cat}}{k_1}$

The v is the initial velocity and must be measured early in the reaction, before product has built up. The V_{max} is the velocity when all enzyme-active sites are filled with substrate. This is theoretical because it can never really happen, some free enzymes will always be around. K_m, the Michaelis constant, is numerically equal to the substrate concentration that gives half of the maximal velocity. This can be estimated from a Michaelis–Menten graph (Figure 8.3). By estimating the V_{max}, you can calculate what half of V_{max} will be, go over to the curve, and drop down to the concentration of substrate, where [S] gives $V_{max}/2$ is the K_m.

V_{max} and K_m are usually referred to as constants and are the first two quantities determined for an enzyme-catalyzed reaction. Note that V_{max} is not a constant at all because it depends on the total [E] and is therefore *only* a constant for your particular experiment. K_m is a constant, however, at least under given conditions of pH, ionic strength, and choice of substrate. If we increase [E], we will increase the velocity at any [S], but the [S] that gives $V_{max}/2$ will *not* change.

FIGURE 8.3 *Determination of K_m and V_{max} with a Michaelis–Menten graph*

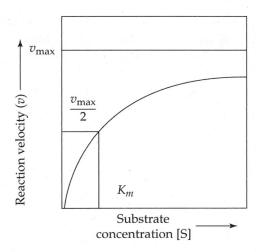

8.4 Significance of K_m and V_{max}

K_m is important for several reasons. We have said that it tells the [S] needed to reach $V_{max}/2$. That in itself may not seem terribly important, but it allows us to better plan our experiments. If you want to test the effect of [S] on v, then you will choose a concentration of S that is close to K_m, where a small change will affect the rate (see Figures 8.2 or 8.3). However, if you want to measure the quantity of enzyme in a system, then you want [E] to be the only variable affecting the rate. Therefore, you choose an [S] that is very high so that there is no effect of [S] on v. For such a test, you would use [S] of 5 to 10 times the K_m.

K_m is also important because it gives an idea of the affinity between the enzyme and substrate. $K_m = (k_2 + k_{cat})/k_1$, or the rate of breakdown of ES divided by the rate of formation of ES. Often, the catalytic rate constant k_p is much slower than the individual rate constants for formation and breakdown, k_1 and k_2. Under those circumstances, K_m is the same as K_s, a true dissociation constant, and a low K_m means that the ES complex forms rapidly. Enzymes often have multiple substrates. A substrate with a lower K_m will bind more quickly to the enzyme than one with a high K_m.

Suppose you isolate a new enzyme from human liver and determine that the K_m for the substrate is 5 mM. The naturally occurring level of that substrate in the liver, however, is 5 nM. You then conclude one of the following: (1) The enzyme was damaged, (2) the reaction does not happen quickly, or (3) the substrate you isolated was not the true or best substrate.

In a clinical lab, blood or other samples are often screened for various enzymes. Sometimes the total enzyme level is measured, in which an increase or decrease may indicate a disease state. Other times, K_m is measured because an altered K_m may indicate a damaged enzyme or another form of the enzyme that should not be present.

V_{max} is also important, although it is not a constant. The basic rate law is

$$v = k_{cat}[ES]$$

The constant k_{cat} is the catalytic rate constant, and the enzyme and substrate must be in the ES form to react. As noted earlier, the theoretical V_{max} is achieved if all enzyme is in the ES form, so

$$V_{max} = k_{cat}[E_t]$$

where E_t is the total concentration of enzyme. Rearranging gives

$$k_{cat} = \frac{V_{max}}{[E_t]}$$

By calculating V_{max} and the actual molar concentration of enzyme, we can calculate k_{cat}. This constant is also called the **turnover number,** which is the number of substrate molecules transformed to product per unit time by one enzyme molecule under maximal conditions. This is a measure of the speed and efficiency of an enzyme.

8.5 Linear Plots

We can estimate K_m and V_{max} from a curve similar to that shown in Figure 8.2, but there are problems with this. First, because V_{max} is never really attained, we can't really measure it. Second, because we plot K_m as $V_{max}/2$, it too has error. Luckily, many investigators have manipulated the Michaelis–Menten equation to give a linear plot. The most common is the well-known Lineweaver–Burk plot shown in Figure 8.4. If we rearrange the Michaelis–Menten equation and take reciprocals, we get the equation below:

$$\frac{1}{v} = \frac{K_m}{V_{max}[S]} + \frac{1}{V_{max}}$$

This equation can be graphed as a straight line. Recall the equation for a line is $y = mx + b$. With the Lineweaver–Burk equation, y is $1/v$. The x value is

FIGURE 8.4 *Lineweaver–Burk plot*

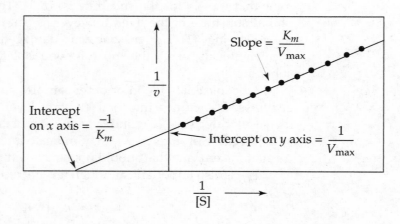

1/[S]. The slope of the line, *m*, is K_m/V_{max}, and the *y*-intercept is $1/V_{max}$ when the line is extended to the *x* axis, it intersects at $-1/K_m$.

PRACTICE SESSION 8.1

Velocity vs. [S] Data

Velocity (μmol/min)	[S] (M)
130	6.5×10^{-4}
116	2.3×10^{-4}
87	7.9×10^{-5}
63	3.9×10^{-5}
30	1.3×10^{-5}
10	3.7×10^{-6}

Using the data given in the adjacent table, calculate K_m and V_{max}. Calculate the turnover number, assuming that you used 0.1 mL of a 0.3 mg/mL solution of enzyme with a molecular weight of 136,000.

First, change the data to reciprocals so you can use the Lineweaver–Burk plot.

Then plot the values on a graph as shown in Figure 8.5. From the graph, the *y* axis = $1/V_{max}$ = 0.006.

$$V_{max} = \frac{1}{0.006} = 166.6 \ \mu mol/min$$

From the *x* axis, $-1/K_m = 2.06 \times 10^4 \ M^{-1}$

$$K_m = 4.85 \times 10^{-5} \ M$$

To calculate the turnover number, you need to determine how many moles of enzyme that you used. The molecular weight of the enzyme is given as 136,000. You used 0.1 mL of a 0.3 mg/mL solution, so

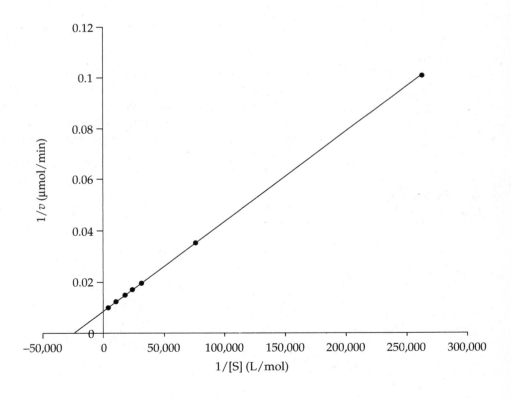

FIGURE 8.5 *Lineweaver–Burk plot for Practice Session 8.1*

Reciprocal Data

1/v (μmol/min)	1/[S] (L/mol)
0.00769	1.539×10^3
0.00862	4.35×10^3
0.0115	1.27×10^4
0.0159	2.56×10^4
0.0333	7.69×10^4
0.1000	2.7×10^5

$$0.3\,\text{mg/mL} \times 0.1\,\text{mL} = 0.03\,\text{mg of enzyme}$$

$$\frac{0.03\,\text{mg}}{136,000\,\text{mg/mmol}} = 2.2 \times 10^{-7}\,\text{mmol} = 2.2 \times 10^{-4}\,\mu\text{mol}$$

$$\text{Turnover number} = k_{cat} = \frac{V_{max}}{\text{enzyme}} = \frac{166.6\,\mu\text{mol/min}}{2.2 \times 10^{-4}\,\mu\text{mol}}$$

$$= 757,273/\text{min}$$

Alternative Plots

For those who have done the Lineweaver–Burk plot before or those who recognize its inherent inferiority, many other plots are more popular nowadays. The Eadie–Hofstee plot uses the equation below:

$$v = -K_m \frac{v}{[S]} + V_{max}$$

Velocity is plotted versus v/[S]. The y intercept gives V_{max}; the x axis gives V_{max}/K_m, and the slope is $-K_m$. This plot is considered superior to the Lineweaver–Burk plot because it has only one reciprocal, which reduces errors, and the substrate concentrations are weighted equally. In addition, the points do not cluster around the y axis. Figure 8.6 shows an Eadie-Hofstee plot.

The Eisenthal–Cornish–Bowden plot uses the equation shown below:

$$V_{max} = v + K_m \frac{v}{[S]}$$

With this plot, however, the points are not plotted as on a normal graph. The coordinates for v and [S] are plotted directly on their axes, and then the points are connected. Where the lines intersect gives V_{max} on the y axis and K_m on the x axis as shown in Figure 8.7. *This is the easiest and one of the most accurate plots (by hand) for quickly determining* K_m *and* V_{max}.

FIGURE 8.6 *Eadie–Hofstee plot*

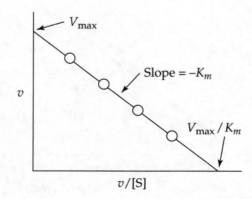

FIGURE 8.7 *Eisenthal–Cornish–Bowden plot*

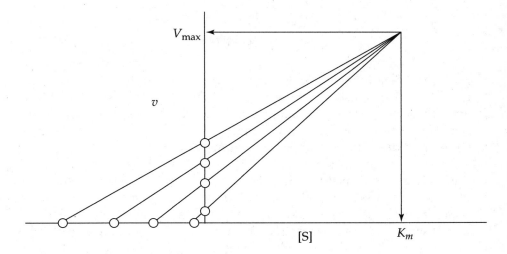

8.6 Properties of Tyrosinase

Tyrosinase, also called polyphenol oxidase, is present in plant and animal cells. In mammalian cells, it catalyzes two steps in the synthesis of melanin pigments from tyrosine. It is located in the skin and is activated by UV sunlight, leading ultimately to a suntan.

Mushroom tyrosinase is tetrameric with a total MW of 128,000. Four Cu^+ atoms are associated with the active enzyme. One of the natural substrates of tyrosinase is 3,4-dihydroxyphenylalanine (DOPA). To study the kinetics of tyrosinase, we will use the reaction shown in Figure 8.8 and the fact that the product, dopachrome, absorbs at 475 nm.

8.7 Why Is This Important?

To understand the nature of metabolism, we must understand the nature of the enzymes that catalyze the reactions. Enzyme kinetics is the study of reaction rates of enzyme-catalyzed reactions. By observing how reaction rates are affected by the concentration of enzyme, substrate, inhibitors, and activators, we will begin to understand the nature of the reaction. For example, if we look at the K_m's for two enzymes that catalyze the formation of glucose 6-phosphate from glucose, hexokinase, and glucokinase,

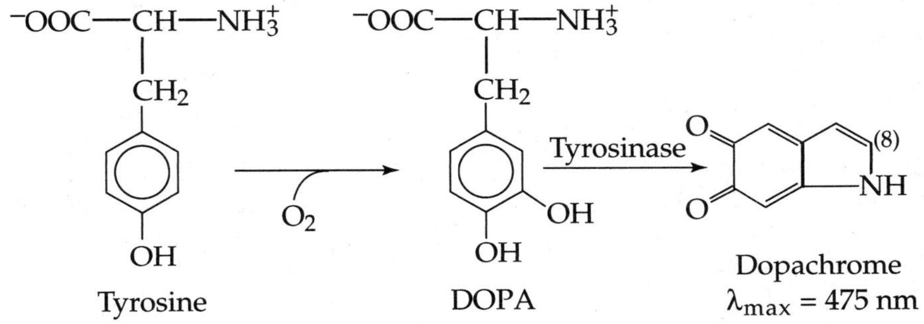

FIGURE 8.8 *The tyrosinase reaction*

$$^-OOC-CH-NH_3^+ \quad\quad ^-OOC-CH-NH_3^+$$

Tyrosine $\xrightarrow[O_2]{}$ DOPA $\xrightarrow{\text{Tyrosinase}}$ Dopachrome λ_{max} = 475 nm

ESSENTIAL INFORMATION

The kinetics of an enzyme-catalyzed reaction can tell us a lot about the nature of the enzyme and the metabolism involved. It was discovered long ago that the relationship of the reaction rate to the amount of enzyme was different from the relationship of the reaction rate to the amount of substrate. This difference led to our current model of simple enzyme-catalyzed reactions. The reaction rate is linearly dependent on the amount of enzyme. Therefore, always be careful when you pipet the enzyme. For nonallosteric enzymes, the rate is hyperbolic to the substrate concentration. The rate increases quickly with increasing substrate at low substrate levels but does not increase much at all with the same level of increase if the substrate level is already high. When comparing amounts of enzymes between different samples, use high concentrations of substrate. In this way, the only variable is the enzyme itself. When studying the effect of substrate on enzymes, use a broad range of substrate levels, from low to very high. Two parameters that are often calculated are K_m and V_{max}. V_{max} is the maximum rate that occurs with infinite substrate concentration. K_m is the concentration of substrate that gives one half the rate of V_{max}. Many types of plots are available to determine these parameters. The most common linear plot is the Lineweaver–Burk plot. The simplest and fastest is the Eisenthal–Cornish–Bowden plot.

we can tell something about the importance of these two enzymes. The K_m for glucose is 0.15 mM with hexokinase and 20 mM with glucokinase. Glucokinase is in the liver only, whereas almost all cells have hexokinase. The normal level of blood glucose is 5 mM, so we can predict that hexokinase will utilize glucose well in all tissues. Glucokinase, on the other hand, only processes glucose when the blood glucose is higher, such as after a carbohydrate-rich meal. This means that when glucose is low, other tissues have a better opportunity to use it. Only when glucose is high will the liver use its other enzyme, glucokinase, to help process it.

Experiment 8

Enzyme Kinetics of Tyrosinase

In this experiment, you will study enzyme kinetics using mushroom tyrosinase. Enzyme and substrate will be varied in separate parts to demonstrate the different nature of the effects of each on reaction velocity. The kinetic parameters, K_m, V_{max}, and k_{cat} will be determined.

Prelab Questions

1. For the first part of the experiment, in which you are varying the amount of tyrosinase, how many milliliters of 15 mM L-DOPA will be in each tube?

2. If I give you two tubes and tell you that one is enzyme and the other substrate, what experiment will you do to determine which is which? For this, you have no other source of enzyme or substrate to use, but you do have spec tubes and a spectrophotometer.

Objectives

Upon successful completion of this lab, you will be able to

- Observe and analyze effects of changing enzyme concentration on initial reaction rates.
- Observe and analyze the effect of changing substrate concentration on initial reaction rates.
- Construct v versus [S] plots and estimate K_m and V_{max}.
- Construct Lineweaver–Burk plots and determine K_m and V_{max}.
- Calculate k_{cat}, given total enzyme concentrations.

TIP 8.1 The beauty of this type of experiment is that each tube is its own reagent blank because the absorbencies are changing. Therefore, you don't have to limit yourself to just one cuvette.

- Design protocols for determining kinetic parameters.
- Explain the significance of K_m, V_{max}, and k_{cat}.

Experimental Procedures

Materials

Sodium phosphate buffer, 0.1 M, pH 7.0

Mushroom tyrosinase (0.2 mg/mL)

L-DOPA, 15 mM

Spectrophotometers

Cuvettes

Methods

Part A: Determining a Tyrosinase Level for Kinetic Assays

In this part of the experiment, you will determine the proper amount of tyrosinase to use for the kinetic analyses. If too little enzyme is used, then the change in A_{475} will be too small to detect, especially at low substrate. On the other hand, if too much is used, the substrate will be depleted too quickly, and the rate will not be linear for a measurable time. For the dopachrome assay, you want a linear rate for at least 1 min while taking 15-s-interval (or less) time points.

1. Set up a protocol for the determination of the correct tyrosinase concentration. Each tube should have a total of 3 mL of solution and be 5 mM in L-DOPA. The volume will be controlled by the phosphate buffer, and the amount of enzyme will vary. Do at least five tubes ranging from 0.1 to 0.5 mL of tyrosinase.

2. Pipet the L-DOPA and phosphate buffers.

3. Pipet the chosen amount of enzyme into one tube. Mix by inversion and immediately read the change in absorbance at 475 nm.

4. Measure for 2 min, recording at 15-s intervals. Repeat steps 2–4 for the other enzyme concentrations. *Remember: Add the enzyme immediately before reading each tube!*

5. Determine the rate for five different tyrosinase levels. The final amount chosen for the next part should give you a change in absorbance per minute ($\Delta A/\Delta$ min) between 0.2 and 0.3/min. The levels given in step 1 are only suggestions. You may have to try higher or lower volumes.

Part B: Determining Kinetic Constants of Tyrosinase

Now that the appropriate enzyme level has been determined, the kinetic parameters, K_m, V_{max}, and k_{cat} will be determined. In Part A, the L-DOPA was saturating. In this part, nonsaturating levels will be used.

1. Set up a protocol as before using the level of tyrosinase that you chose in step 5 of Part A. There should still be 3 mL per tube. The recommended levels of L-DOPA are 0.05, 0.10, 0.20, 0.40, 0.80, and 1.0 mL.

2. Follow the same procedures for addition, mixing, and recording as in Part A.

Analysis of Results

Experiment 8: **Enzyme Kinetics of Tyrosinase**

Data

Part A

In the following table, indicate the absorbencies at the time points you used for the varying enzyme volumes.

Enzyme Volume (mL)	Start	15″	30″	45″	60″	75″	90″	105″
0.1								
0.2								
0.3								
0.4								
0.5								

Part B

In the following table, indicate the absorbencies recorded for the time points you used for the varying substrate levels.

Substrate Volume (mL)	Start	15"	30"	45"	60"	75"	90"	105"
0.05								
0.10								
0.2								
0.4								
0.8								
1.0								

Calculations

The concentration of the tyrosinase is 0.2 mg/mL. The extinction coefficient for dopachrome at 475 nm is $3600 \, M^{-1} cm^{-1}$.

Part A

1. Calculate the rate of reactions as $\Delta A / \Delta$ min. You may want to graph the absorbencies versus time to establish the initial velocity. If so, turn in your graphs with this write up. If you choose to calculate it otherwise, show how you calculated the rate.

2. Plot rate (μmol/min) versus E in milligrams. The changes in absorbance per minute must be converted to micromoles, using Beer's law and the extinction coefficient for dopachrome. This is similar to the calculations we did earlier for LDH:

$$\mu\text{mol product/min} = \frac{(\Delta A / \Delta \text{ min})}{3600 \, M^{-1}} \times (10^6 \, \mu M/M) \times 0.003 \, L$$

Enzyme Volume (mL)	ΔA_{475}/min	μmol/min
0.1		
0.2		
0.3		
0.4		
0.5		

What is the amount of enzyme that you chose to use in Part B?

Part B

1. Calculate the units of enzyme activity for the varying amounts of substrate. Calculate the substrate concentrations in the test tubes at time = zero point.

Substrate Volume (mL)	L-DOPA (mM)	ΔA_{475}/min	μmol/min
0.05			
0.1			
0.2			
0.4			
0.8			
1.0			

2. Make a Michaelis–Menten graph. Plot V (μmol/min) versus [S] (mM). Estimate K_m and V_{max} from this graph.

   K_m from Michaelis–Menten graph   _____

   V_{max} from Michaelis–Menten graph   _____

3. Make any *linear* plot (L-B, ECB, EH, and so on) to determine K_m and V_{max}.

   K_m from linear plot   _____

   V_{max} from linear plot   _____

4. Use the molecular weight of tyrosinase, 128,000, to determine how many micromoles of tyrosinase are in the tubes that you used to calculate V_{max}.

5. The turnover number is the number of moles of product produced per minute by 1 mol of enzyme at V_{max}. Using the number you calculated in item 4 and the V_{max} from your linear graph, calculate the turnover number for tyrosinase.

Questions

1. A competitive inhibitor competes with substrate for binding to the active site of the enzyme. The enzyme, once bound by the inhibitor, cannot form product. How does a competitive inhibitor affect the velocity of product formation? Do you need more or less of the substrate to get the same velocity as found before the inhibitor was added?

2. Suggest a way that a competitive inhibitor can be used as a drug against a disease.

3. Enzyme X has a molecular weight of 48,000. It converts substrate Z into product Y. Z absorbs at 340 nm, and Y absorbs at 480 nm.

   a. At what wavelength do you measure the change in absorbance to assay for enzyme X? Does the absorbance increase or decrease over time?

   b. If $V_{max} = 60\,\mu$mol/min and you use 400 μL of a 0.1 mg/mL solution of enzyme, what is the turnover number?

4. Why is V_{max} *not* a constant? Why do we want to analyze k_{cat} instead of V_{max}?

Experiment 8a

Enzyme Kinetics of LDH

In this experiment, you will study the kinetic parameters of your LDH. You will use linear plots to determine the K_m for NAD^+ and the V_{max}. You will then calculate an estimated turnover number, k_{cat}.

Objectives

Upon successful completion of this lab, you will be able to

- Observe and analyze effects of changing enzyme concentration on initial reaction rates.

- Observe and analyze the effect of changing substrate concentration on initial reaction rates.

- Construct v versus [S] plots and estimate K_m and V_{max}.

- Construct Lineweaver–Burk plots and determine K_m and V_{max}.

- Calculate k_{cat}, given total enzyme concentrations.

- Design protocols for determining kinetic parameters.

- Explain the significance of K_m, V_{max}, and k_{cat}.

Experimental Procedures

Materials

CAPS buffer, 0.15 M, pH 10

NAD^+, 6 mM

Lactate, 0.15 M

Spectrophotometers

Cuvettes

TIP 8.1 The beauty of this type of experiment is that each tube is its own reagent blank because the absorbencies are changing. Therefore, you don't have to limit yourself to just one cuvette.

TIP 8.2   *Never* vortex enzyme solutions! Always   cover the top of the tube.
mix by quick but gentle inversion using Parafilm to

Procedure

1. Using your best sample of LDH and the standard assay, vary the amount of LDH and calculate the initial velocity for five different volumes. Find an amount that gives a change in absorbance versus time of about 0.2/min using the standard assay. You may have to experiment with this to get it. The acceptable range is 0.15–0.25/min.

2. Once you find the correct amount of LDH to use, hold this value constant for the duration of this experiment.

3. Using the standard NAD^+ solution, adjust the assay so that you add progressively less NAD^+ each time. Add water to compensate for the lost NAD^+ volume.

4. Record the absorbance changes over time and calculate the $\Delta A/\Delta$ min. If you get down to 50 μL of the NAD^+ and still are not seeing much decrease in activity, make a dilution of the NAD^+ and try again.

5. When you have assayed volumes of NAD^+ that run a range from full activity down to almost no activity and this includes at least six good assays, you probably have enough data to analyze.

6. Calculate the concentration of NAD^+ in each cuvette at the beginning of the assay and the initial activity recorded in micromoles/minute.

7. Use your favorite linear graph to calculate K_m for NAD^+ and V_{max} for the amount of enzyme you chose.

8. Using the known protein concentration of the LDH sample that you used and assuming (falsely) that all the protein is LDH, calculate the k_{cat} for LDH, assuming a molecular weight of 150,000.

Name _____

Section _____

Lab partner(s) _____

Date _____

Analysis of Results

Experiment 8a: **Enzyme Kinetics of LDH**

Data

1. In the following table, indicate the absorbencies at the time points you used for the varying enzyme volumes.

Enzyme Volume (mL)	Start	15"	30"	45"	60"	75"	90"	105"

2. In the following table, indicate the absorbencies recorded for the time points you used for the varying substrate levels.

Substrate Volume (mL)	Start	15″	30″	45″	60″	75″	90″	105″

Calculations

1. Calculate the rate of reactions as $\Delta A / \Delta$ min. You may want to graph the absorbencies versus time to establish the initial velocity. If so, turn in your graphs with this write up. If you choose to calculate it otherwise, show how you calculated the rate.

2. Plot rate (μmol/min) versus E in milliliters. The changes in absorbance per minute must be converted to change in micromoles, using Beer's law and the extinction coefficient for NADH.

$$\mu\text{mol product/min} = \frac{(\Delta A / \Delta \text{ min})}{6220 \text{ M}^{-1}} \times (10^6 \, \mu\text{M/M}) \times 0.003 \text{ L}$$

Enzyme Volume (mL)	Δ A$_{340}$/min	μmol/min

What is the amount of enzyme that you chose to use for the second part with varying substrate?

3. Calculate the units of enzyme activity for the varying amounts of substrate. Calculate the substrate concentrations in the test tubes at time = zero point.

Substrate Volume (mL)	mM NAD$^+$	Δ A$_{340}$/min	μmol/min

4. Make a Michaelis–Menten graph. Plot v (μmol/min) versus [S] (mM). Estimate K_m and V_{max} from this graph.

    K_m from Michaelis–Menten graph _____

    V_{max} from Michaelis–Menten graph _____

5. Make any *linear* plot (L-B, ECB, EH, and so on) to determine K_m and V_{max}.

    K_m from linear plot _____

    V_{max} from linear plot _____

6. Use the molecular weight of LDH, 150,000, and the protein concentration that you calculated previously to determine how many micromoles of LDH are in the tubes that you used to calculate V_{max}.

7. The turnover number is the number of moles of product produced per minute by 1 mol of enzyme at V_{max}. Using the number you calculated in item 6 and the V_{max} from your linear graph, calculate the turnover number for LDH.

8. If you want to determine the K_m for lactate, what protocol do you set up?

9. Why does the clever lab instructor choose to have you determine the K_m for NAD^+ instead of for lactate?

10. Which of the components of the assay for K_m of NAD^+ have the least effect on your results if they are pipetted imprecisely? Why?

Additional Problem Set

1. For the hypothetical reaction

$$3A + 2B \rightarrow 2C + 3D$$

the rate was experimentally determined to be

$$rate = k[A]^1[B]^1$$

What is the order of the reaction with respect to A? With respect to B? What is the overall order of the reaction?

2. For an enzyme that displays Michaelis–Menten kinetics, what is the reaction velocity v (as a percentage of V_{max}), observed at (a) $[S] = K_m$, (b) $[S] = 0.5\,K_m$, (c) $[S] = 0.1\,K_m$, (d) $[S] = 2\,K_m$, and (e) $[S] = 10\,K_m$?

3. How is the turnover number related to V_{max}?

4. Why is V_{max} *not* a true constant?

5. Determine the K_m and V_{max} for an enzymatic reaction, given the data shown in the adjacent table.

6. Why do we do linear transformations of the Michaelis–Menten equation?

7. A standard line fits the equation $y = mx + b$, where m is the slope and b is the y intercept. What are these values for (a) a Lineweaver–Burk plot, (b) an Eadie–Hofstee plot, (c) an Eisenthal–Cornish–Bowden plot?

Substrate Concentration (mM)	Velocity (μmol/min)
2.500	0.588
1.000	0.500
0.714	0.417
0.526	0.370
0.250	0.256

Webconnections

For a list of websites related to the material covered in this chapter, go to **Webconnections** at the *Experiments in Biochemistry* site on the Brooks/Cole Publishing website. You can access this page at http://www.brookscole.com and follow the links from the chemistry page.

References and Further Reading

Boyer, R. F. *Modern Experimental Biochemistry.* Menlo Park, CA: Addison-Wesley, 1993.

Campbell, M. *Biochemistry.* Philadelphia: Saunders, 1998.

Cornish-Bowden, A. *Analysis of Enzyme Kinetic Data.* New York: Oxford University Press, 1995.

Dryer, R. L., and G. F. Lata. *Experimental Biochemistry.* New York: Oxford University Press, 1989.

Gutfreund, H. *Kinetics for Life Sciences: Receptors, Transmitters, and Catalysts.* Cambridge: Cambridge University Press, 1995.

Jack, R. C. *Basic Biochemical Laboratory Procedures and Computing.* New York: Oxford University Press, 1995.

Leatherbarrow, R. J. "Linear and Non-linear Regression of Biochemical Data." *Trends in Biological Science* 15, 1990.

Purich, D. L., J. N. Abelson, and M. I. Simon. *Contemporary Enzyme Kinetics and Mechanisms*. San Diego: Academic Press, 1996.

Robyt, J. F., and B. J. White. *Biochemical Techniques*. Long Grove, IL: Waveland Press, 1990.

Schulz, A. R. *Enzyme Kinetics: From Diastase to Multi-enzyme Systems*. Cambridge: Cambridge University Press, 1994.

Segel, I. H. *Biochemical Calculations*, 2nd ed. New York: Wiley Interscience, 1976.

Chapter 9

Electrophoresis

Introduction

In this chapter, we introduce what is probably the most important biochemistry technique for you to learn. Electrophoresis is used to separate biological molecules in an electric field. It is most often used to separate proteins or DNA, and it can be done with a system based on agarose or polyacrylamide. All chapters and experiments that follow rely heavily on electrophoresis.

9.1 Electrophoresis

Electrophoresis is the movement of charged particles in an electric field. A negatively charged particle will move toward the positive pole and vice versa. The velocity at which it moves depends on several factors according to the following equation:

$$v = \frac{qE}{f}$$

where v =   velocity

q =   net charge on the molecule

E =   applied voltage

f =   frictional coefficient

Therefore, a molecule with a charge of -2 will move twice as fast, all else being equal, as a molecule with a charge of -1. Remember that, for most biological molecules, their net charge depends on the medium in which you put them. If the pH of the buffer in the system is changed, the net charges will change on some of the molecules.

If the voltage is increased, the separation will also be quicker, but there are limitations on this. Too high a voltage can destroy the sample or the

TIP 9.1 Anytime you run an electrophoresis experiment, be aware of the following:

1. Are the molecules negatively charged? If not, they will run the wrong way.
2. Is the tracking dye front straight? If not, conditions were not the same in all lanes.
3. If you see a band, how do you know it is what you think it is?
4. What parameter of the molecule causes it to run to that point on the gel? Is it the charge? The size?
5. If you don't see a band, where did it go?
6. Will the bands diffuse quickly or not?

support medium due to excessive heat. Also, the bands are usually sharper and better separated with less streaking if electrophoresis is done slowly. A teaching lab has time limitations, so sometimes less-than-perfect results are accepted so that you can leave at a reasonable hour.

The frictional coefficient is the retarding effect of the size and shape of the particle and the nature of the support medium. A more round protein, for example, will move faster than a rod-shaped one with the same weight. A denser medium retards all proteins but has a greater effect on larger ones.

Although electrophoresis could be used to separate charged molecules from any class of biomolecule, the two most common are proteins and nucleic acids. In this chapter, we discuss separation of proteins, and Chapter 11 deals with DNA separation.

Proteins are different from one another based on their amino acid sequence. The amino acid sequence gives each protein a unique charge character as well as size and shape. All these factors act together to affect how the proteins migrate with electrophoresis.

Many different media can be used for electrophoresis, such as liquid, paper, or gel. Most electrophoresis done today uses a gel-based medium.

9.2 Agarose Gels

Agarose is one of the most common supports for electrophoresis. Agarose is a natural polysaccharide of galactose and 3,6-anhydrogalactose derived from agar, which is itself obtained from certain red algae. The repeating unit is shown in Figure 9.1.

FIGURE 9.1 *The structure of agarose*

Agarose

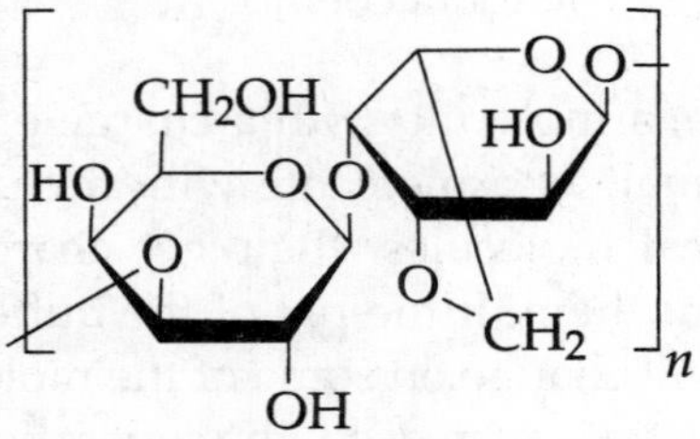

3,6-anhydro bridge

TIP 9.2 Weigh out your agarose carefully because a huge difference exists between a 0.8% gel and a 1% gel. Be sure not to let your agarose boil too long, or it will become too concentrated.

Agarose chains tend to make left-handed helices that intertwine with each other. This gives rise to a gel that is quite dense for its concentration. A solution of only 1.0% w/v agarose will solidify into a fairly dense gel that can be used to separate proteins or nucleic acids. Agarose gels are prepared by boiling a defined quantity of the dry polymer in buffer until it melts. The melted agarose suspension is then poured into a casting tray with a well-forming comb and allowed to cool and solidify. This type of gel makes an excellent support for separating proteins based on charge. The spongelike gel allows proteins to pass through quickly but provides enough stability so that they do not diffuse quickly when the power is turned off. It is important to remember that the ions in the buffer carry most of the current during electrophoresis. Because this buffer is also in the gel, some of the current goes through the gel so that the samples can move. If you accidentally make up your gel in water, no current will flow through the gel and your samples will not move.

Agarose gels are usually run horizontally in the submarine mode where the gel is completely immersed in buffer. This aids in heat reduction. Figure 9.2 shows an agarose gel setup. Agarose gels are **native gels** because nothing in the system denatures a protein if everything goes according to plan. Proteins retain their normal conformations and activity, if any, as they run down the gel. However, with an active enzyme, be careful that the temperature does not get too high during the run; otherwise, the enzyme may be denatured enough to destroy its biological activity. Running time, voltage, and buffer volume all control the heat output of electrophoresis.

ESSENTIAL INFORMATION

Agarose gels are native gels in which the molecule of interest retains its native conformation and activity while it runs. Agarose is simple to use. Just heat the correct amount of agarose in a suitable buffer until the agarose melts and then pour the gel. Small differences in concentration are very important. A 0.7% agarose gel is very flimsy, whereas a 1.2% gel is a brick. Always make up agarose gels in a buffer, not water. When proteins separate on an agarose gel, the size, shape, and charge of the protein determine how far it travels in the allotted time. Running the gel at higher voltage will speed the process but reduce the quality of your results.

FIGURE 9.2 *Submarine gel setup*

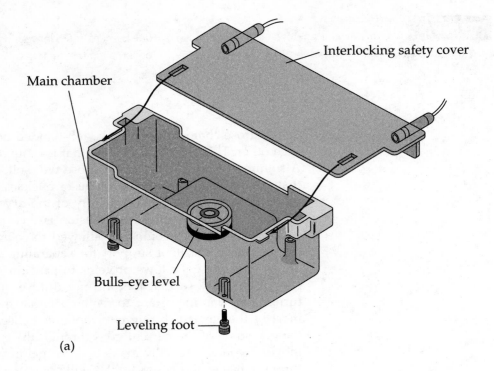

Main chamber

Interlocking safety cover

Bulls-eye level

Leveling foot

(a)

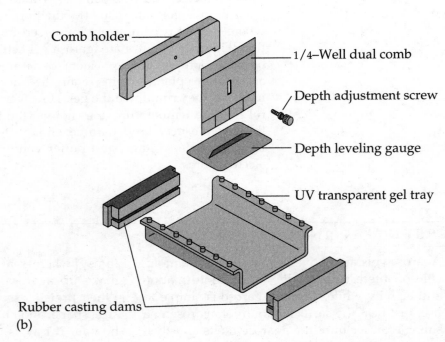

Comb holder

1/4–Well dual comb

Depth adjustment screw

Depth leveling gauge

UV transparent gel tray

Rubber casting dams

(b)

9.3 Polyacrylamide Gels

Polyacrylamide gels are long polymers of acrylamide cross-linked with *N,N′*-methylenebisacrylamide, as shown in Figure 9.3. Unlike the simple agarose gels, a polyacrylamide gel has many components, and the nature of the gel is controlled by the amounts chosen.

FIGURE 9.3 *The structure of cross-linked polyacrylamide*

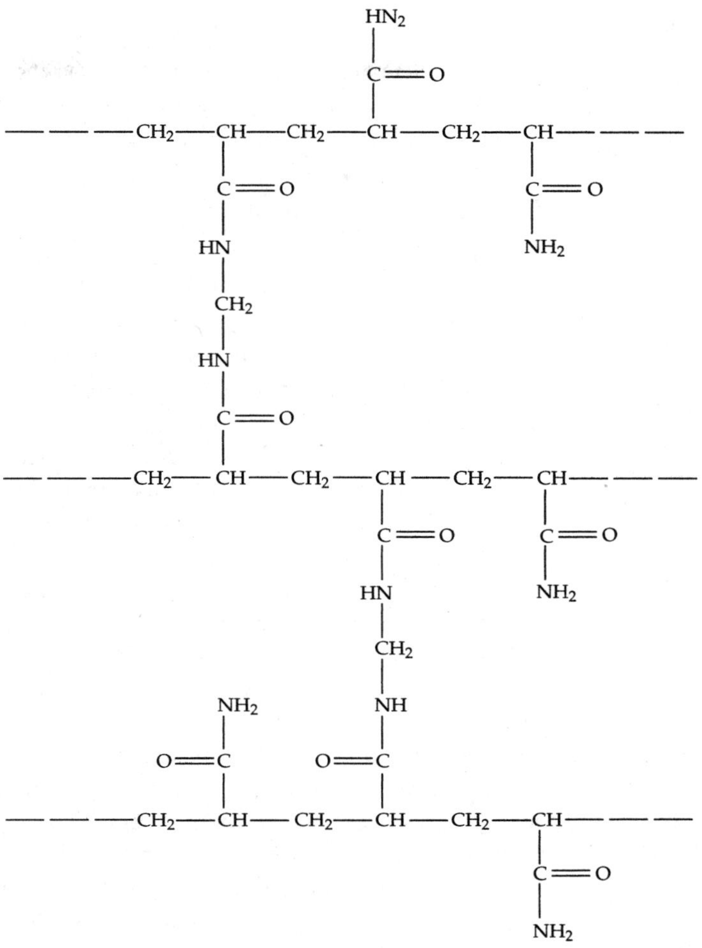

Total Concentration of Acrylamide

The mobility, μ, for an electrophoresis is the ratio of the velocity the particle moves divided by the voltage applied. The following equation describes the relationship of mobility to acrylamide concentration:

$$\mu = e^{-kT}$$

where k is a constant for a certain percentage of bisacrylamide and T is the total acrylamide concentration. Therefore, the higher the concentration, the slower particles move.

Amount of Bisacrylamide Cross-Linker

This is not as important as you might think. Most students assume that, to control the pore size of a gel, the cross-linking reagent is the most important. However, a cross-linker has an optimum percentage, which is between 3 and 5% of the total, which gives the best gels.

Temed

TEMED is a catalyst that stimulates the formation of free radicals during the reaction that links acrylamide molecules. The amount of TEMED controls the speed at which the gel will harden.

Ammonium Persulfate

Ammonium persulfate, the initiator of the reaction, creates free radicals and begins a chain reaction that links all the acrylamide molecules together. Like TEMED, the amount of ammonium persulfate controls the speed. Small volumes of highly concentrated ammonium persulfate are usually used. Weigh and pipet it carefully. A small error can be the difference between the gel not solidifying during this geological epoch and it solidifying in the flask before you can pour it between the plates.

Many times a discontinuous gel is made where the bulk of the gel is high percentage at pH 8.5, but a couple of centimeters of gel on top is low percentage (3%) at pH 6.5. This upper gel is the **stacking gel** because it tends to compress all of the proteins into a thin band. The lower gel is called the **running, resolving,** or **separating gel** because the proteins in it separate from each other with the small proteins running fastest. The stacking gel works in two ways. First, because it is a low-concentration gel, proteins move quickly in it. When the proteins encounter the higher-density separating gel, they slow down. Therefore, proteins that enter the separating gel first are going slower than the ones still in the stacking gel. This causes an accordion effect, and the protein sample becomes much thinner as it enters the separating gel. Second, the pH difference plays upon the charge nature of the compounds being used. Glycine, which is in the buffer, has only a partial negative charge at pH 6.5. This causes zones to be set up in the lane where proteins are sandwiched between chloride ions with lots of negative charge and glycine ions with less charge. This causes voltage differences within the zones that tend to push all proteins together. Once proteins and glycine enter the separating gel, the pH increase to 8.6 puts more negative charge on the glycine, relieving this effect. Proteins will then move at different rates based on their size.

Many different apparatuses are available for running polyacrylamide gels. A common one is the Bio Rad Mini Protean II shown in Figure 9.4.

9.4 SDS-PAGE

As we know, electrophoresis separates proteins based on size, shape, and charge. Native gels are often difficult to interpret because of these three variables. When using electrophoresis to determine the molecular weight of a protein, electrophoresis is usually done in the presence of the detergent sodium dodecyl sulfate (SDS), which has the structure

$$CH_3-(CH_2)_{11}-SO_4^-$$

SDS binds to proteins in a constant ratio of 1.4 g of SDS per gram of protein and covers the protein with negative charges. SDS and some

FIGURE 9.4 *Bio Rad Mini Protean II electrophoresis system*
(Courtesy of Bio Rad)

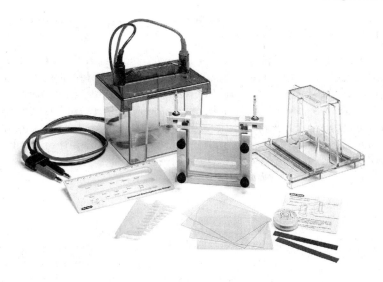

β-mercaptoethanol included in the sample buffer also denature proteins and break up any polymers into their subunits. The β-mercaptoethanol reduces any disulfide bridges present. The effect is that all proteins attain the same shape (random coil) and have the same charge-to-mass ratio. The only variable left is the mass. Proteins therefore separate on the gels solely based on molecular weight.

SDS-PAGE (polyacrylamide gel electrophoresis) is often used to determine if a protein is pure and if subunits are present. If LDH (see Section 9.6) is run on an SDS gel, the LDH would be broken into its subunits. Instead of having a native tetramer migrating down the gel, the individual M and H monomers will migrate. For example, you might run a protein on gel filtration and see a molecular weight (MW) of 100,000. Then if you run the same protein on an SDS-PAGE, you might see one band with a molecular weight of 50,000. This tells you that your protein is made up of two equal-sized subunits. On the SDS-PAGE, you might see two bands, with one at MW 75,000 and the other at MW 25,000. Because these add up to MW 100,000, you conclude that you have two unequal-sized subunits in the native molecule.

Standard Curves for Molecular Weight

One of the main uses of SDS-PAGE is the determination of molecular weight. When gels are stained and destained, blue bands show up at locations on the gel based on molecular weight. If the log of the molecular weight is graphed versus the R_m of the protein band (distance the band traveled/distance the

TIP 9.3 Remember that SDS-PAGE separates proteins based on their size (molecular weight). It also breaks up proteins into subunits. The number that you calculate will be the molecular weight of subunits, if any.

tracking dye traveled), a straight line is seen for most proteins. Usually, several protein standards are run along with an unknown. The molecular weight of the unknown is calculated by interpolation from the standard curve. Remember to use two-cycle log paper if your standards have molecular weights that span more than one order of magnitude. Most computer graphing programs can also plot log MW for you. Figure 9.5 shows a typical calibration curve for proteins of known molecular weight with an SDS-PAGE.

9.5 Staining Gels

Many different dyes are used during electrophoresis, and it is important (and often confusing) to keep their uses straight. The first type is bromophenol blue, which is included in the sample buffer (also called the tracking dye or loading dye). This acts as a marker so that you can see how the separation is proceeding. During the run, it is the only band that you will see. Bromophenol blue is negatively charged and small, so it migrates more quickly than the proteins that you are trying to separate. By using bromophenol blue, you can be confident that if the dye has not run off the gel, then neither have your proteins.

The second is Coomassie Blue, similar to the dye used in the Bradford protein assay. Used after an electrophoresis is over, it stains all proteins blue and shows where the proteins are on the gel. Figure 9.6 shows a typical result of a Coomassie gel.

Other stains are available, such as copper stain or silver stain. They often give more sensitivity and can therefore be used with smaller amounts of protein on the gel.

Another way to stain a gel is with a chemical specific for a particular protein or enzyme. This is called an activity stain. In Experiment 9, we will use an activity stain that is specific for lactate dehydrogenase. It contains NAD^+, lactate, phenazine methosulfate (PMS_{ox}), and nitroblue tetrazolium (NBT_{ox}), which undergo the following reaction:

$$NAD^+ + lactate \rightarrow NADH + pyruvate$$
$$NADH + PMS_{ox} \rightarrow NAD^+ + PMS_{red}$$
$$PMS_{red} + NBT_{ox} \rightarrow PMS_{ox} + NBT_{red}$$

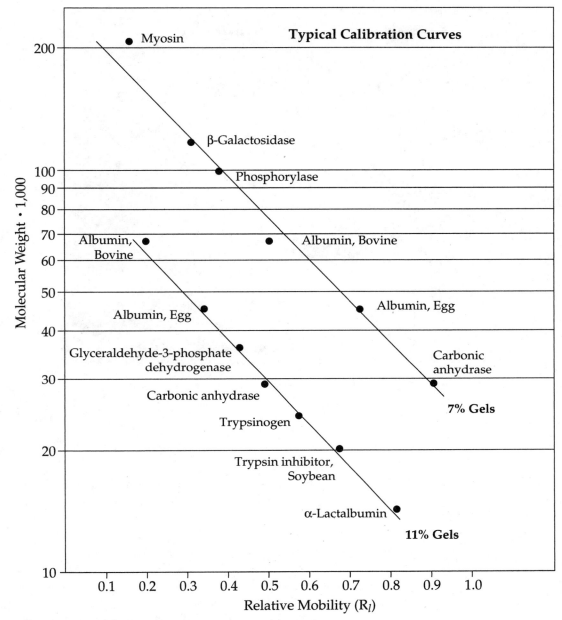

FIGURE 9.5 *SDS-PAGE calibration curves*

The first reaction occurs only if lactate dehydrogenase is present. Reduced NBT forms a purple precipitate, so wherever LDH is present in the gel, there will be a purple band.

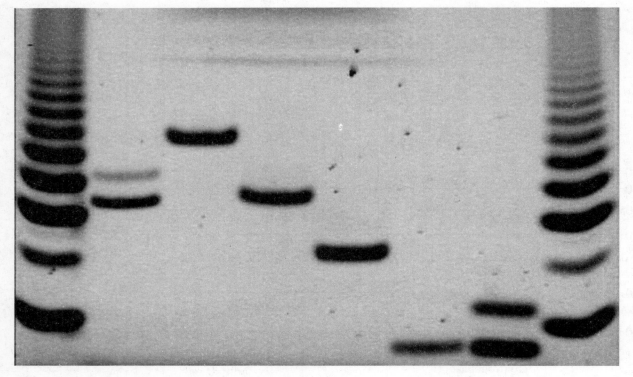

FIGURE 9.6 *Coomassie Blue–stained gel*
(Courtesy of Bio Rad)

9.6 Lactate Dehydrogenase

Lactate dehydrogenase (LDH) is a glycolytic enzyme that is present in all tissues. It catalyzes the final step of anaerobic glycolysis, regenerating NAD^+:

$$\textbf{pyruvic acid + NADH} \rightleftharpoons \textbf{NAD}^+ + \textbf{L-lactic acid}$$

LDH, an allosteric enzyme with four subunits, has a total molecular weight of 150,000. LDH has two types of subunits, H and M, that vary slightly in their amino acid compositions. The H subunit is more prevalent in heart tissue and the M in muscle tissue. These subunits are combined in every possible combination to give four subunits per native LDH molecule (that is, H_4, H_3M, H_2M_2, HM_3, and M_4). Because the two types of subunits have different charges, the five isozymes formed by the different combinations have different migrations on native gels.

The difference in charge between the various isozymes of LDH is the basis of a sensitive clinical test for myocardial infarctions. The heart muscle contains predominantly the H_4 isozyme, with lesser amounts of H_3M. In blood serum under normal circumstances, the predominant isozymes are H_3M, H_2M_2, and HM_3. After a heart attack, damaged heart cells spill their contents into the blood serum, and the blood serum levels of H_4 and H_3M rise dramatically.

Experiments 9–9c comprise the analysis of LDH. In Experiments 9 and 9a, native gel electrophoresis will be used to analyze isozymes of LDH. In Experiments 9b and 9c, SDS-PAGE will be used to determine the purity and subunit molecular weight of LDH.

9.7 Why Is This Important?

This chapter begins our study of electrophoresis, which is perhaps the most used technique outside of spectrophotometry. Modern biochemistry and molecular biology would be impossible without electrophoresis. To understand the results, you must understand the nature of the separation. Was it based on size, shape, or charge or a combination of all three? When seeing bands on a gel, you have to know what they mean. A native gel does not denature the sample. In Experiments 9 and 9a, LDH molecules travel in their native conformations and retain all their activity. This type of electrophoresis is common for purification of proteins and nucleic acids when native conformation is required.

The three biggest fields of study today are protein purification, immunology, and molecular biology. All three use polyacrylamide gels to separate proteins or DNA. A sequencing gel is a large polyacrylamide gel with bands that differ by only one base. The principles for the formation and use of the gel are the same as we will do in Experiments 9b and 9c. We will also use this technique again in Experiment 10 as the first part of the Western Blot.

In addition, if you are going into the sciences, agarose gel electrophoresis and SDS-PAGE are two techniques that you definitely want on your résumé. Knowing these techniques will make you much more marketable once your undergraduate career is over.

9.8 Expanding the Topic

Variables and Controls

Science involves setting up experiments that answer a particular question. In many ways, it is a "black box." You mix some things together, make some measurements, and get some numbers. What these numbers mean depends on how good your experiment was, whether you ran proper controls, and how sure you are that the numbers you generated reflect what you think they do. To get the most out of our experiments, you must do more than just mix stuff together. Ask yourself, "What's the variable?" and "What controls did I run?"

In Experiments 9 and 9a, we will isolate one variable—charge. Agarose is not the best support for a protein of 150,000 daltons (D), which is why the bands will be a little diffused. However, we can get all the information we need because the isozymes separate so nicely from one another. How do we know that we will only separate by charge and that if two bands move different distances they have different charges? We know this because LDH isozymes all have the same size and shape, which are the other variables with proteins. Without understanding this, we will know nothing from the experiment because we have three variables. How do we know what isozymes are present in crude bovine heart LDH? We know because we run controls with LDH 1, 2, and 5. If we run just one LDH sample and see three bands, we will not know if they are LDH 1, 2, and 3 or 2, 3, and 4, and so on.

What's This 6× Stuff?

When a buffer or other reagent is given the abbreviation "something × ," it means that the reagent is that many times more concentrated than it should be in its final form. If we make up a 5× reservoir buffer, then it should be diluted five times before being put into the electrophoresis chamber. If we have a 2× tracking dye (sample buffer), then we mix 1 part sample buffer to 1 part sample before we run it on the gel. This way, the final concentration in the sample well is correct. For electrophoresis tracking dyes, a common designation is 6× because a common sample size is 10 μL; if we add 2 μL of a 6× tracking dye, then we end up with 12 μL of the correct concentration.

Experiment 9

Native Gel Separation of LDH Isozymes (Short Version)

In this experiment, you will use agarose gel electrophoresis to separate several isozymes of LDH and analyze the charge nature of the H and M subunits.

Prelab Questions

1. How much agarose must you weigh out to make the gel?

2. Describe the order of addition of parts and solutions to the casting tray and the order of removal of the parts from the casting trays.

3. Why must the samples be loaded into the wells through a buffer?

Objectives

Upon successful completion of this lab, you will be able to

- Pour and load an agarose gel for electrophoresis.
- Identify LDH isozymes based on their electrophoretic migrations.
- Predict migration patterns for unknown proteins given pI's and molecular weights.

Experimental Procedures

Materials

Electrophoresis chambers and power supplies

Gel trays and combs

Agarose

0.02 M TRIS, 0.02 M glycine, 0.002 M EDTA, pH 8.6 (reservoir buffer)

6× sample buffer with glycerol and bromophenol blue

LDH activity stain, 0.1 M TRIS, 1% lactate, 0.05% NAD$^+$, 0.0005% PMS (phenazine methosulfate) and 0.005% NBT (nitroblue tetrazolium), pH 9.2.

LDH 1, LDH 2, and LDH 5 standards

Crude bovine heart LDH

Crude rabbit muscle LDH

LDH isotrol (LDH 1–5 mixture)

Methods

These procedures were originally written for Fisher Mini-Submarine gels but are essentially the same for all submarine gel apparatuses.

Preparing Agarose

1. Prepare a mixture of 0.8% w/v agarose in reservoir buffer (TRIS-glycine, pH 8.6). Make 40 mL of the solution in a flask.

2. Heat the flask in a microwave oven until the agarose has melted. Swirl to mix. This takes 30–40 s on high power in most microwave ovens if only one flask is present. Do not let the agarose boil too long, or it will boil over.

3. Allow the solution to cool for 2 min before pouring into the casting trays. It should be clear with no significant agarose clumps.

Pouring Agarose

1. Insert the rubber stoppers into the casting tray.

2. Pour 40 mL of agarose into the casting tray.

3. Insert the comb into the gel. Make sure that the orientation is correct to give the right amount of gel on each side of the comb. The comb should not rest against the rubber stopper but should be about 1 cm from it.

4. Let the gel cool for 15 min. Ever so carefully, remove the rubber stoppers before removing the comb. You may have to use a spatula to break the surface tension between the gel and the rubber stopper.

Preparing and Applying the Sample

1. For each LDH sample to be loaded, take 10 μL of sample and add 2 μL of 6× sample buffer. Vortex to mix the contents of the tube. Spin the tubes in a microfuge for a couple of seconds to bring the entire sample back down.

2. Transfer the casting trays to the electrophoresis unit with the sample wells nearest the black (−) electrode.

3. Slowly fill the chamber with reservoir buffer until the buffer is about 1 cm over the top of the gel.

4. Using a Pipetman, apply the samples to the wells. Load the entire sample if it will fit without overflowing but do *not* let the solution overflow into the other wells.

5. Put on the lid and connect the leads to the power supply.

6. Set the voltage to 120 V and electrophorese until the marker dye is about 2 cm from the end.

Staining

1. Obtain a small plastic box with a lid. Carefully slide the gel out of the casting tray and into the box.

2. Cover the gel with about 25 mL of LDH activity stain.

3. Incubate the gel for 15–30 min in a 37°C oven or water bath until the bands develop. Remove the gel from the stain when done and place it in water. Sketch or photograph as directed.

Name _____   Section _____

Lab partner(s) _____   Date _____

Analysis of Results and Questions

Experiment 9: **Native Gel Separation of LDH Isozymes (Short Version)**

Data

Describe the results of your separation by including a sketch or photo of the gel. Be sure to correctly label each lane with the sample that was applied there. Sketch it so that the wells are at the top.

Analysis of Results

1. Why do some lanes have more than one band in them? What does each band represent?

2. Are the isozymes pure?

3. Describe the nature of the LDH molecules that are separated on this system. How do you know that only one variable is at work here?

4. What can you conclude about the pI of the two types of LDH subunits?

Questions

1. Describe the effects the following have on the way your gel is run. Use the following equation to justify your answer:

$$v = \frac{qE}{f}$$

   a. Increasing the agarose concentration to 1.5%

   b. Increasing the pH of the TRIS-glycine buffer to 9.5

   c. Decreasing the voltage at which the gel was run

2. You need to load 10 μg of protein into one of the wells of a gel. This needs to be in 1× buffer and in a total volume of 15 μL. You are given a 10 μg/μL solution of protein, a 5× buffer, and water. How much of each should you mix to load the gel correctly?

Experiment 9a

Native Gel Separation of LDH Isozymes (Comprehensive Version)

In this experiment, you will use agarose gel electrophoresis to separate the isozymes of LDH, determine which ones are in your purified samples, and analyze the charge nature of the H and M subunits.

Objectives

Upon successful completion of this lab, you will be able to

- Pour and load an agarose gel for electrophoresis.
- Identify LDH isozymes based on their electrophoretic migrations.
- Predict migration patterns for the purified LDH isozymes.

Experimental Procedures

Materials

Mini-Submarine gel apparatuses

Power supplies

Agarose

0.02 M TRIS, 0.02 M glycine, 0.002 M EDTA, pH 8.8

6× sample buffer (0.2 M TRIS, 0.2 M glycine, 0.01 M EDTA, 0.02% bromophenol blue, 25% glycerol, pH 8.8)

LDH activity stain, 0.1 M TRIS, 1% lactate, 0.05% NAD^+, 0.0005% PMS, and 0.005% NBT, pH 9.2

LDH 1, LDH 2, and LDH 5 standards

Crude bovine heart LDH

LDH isotrol (contains all five isozymes)

Your LDH samples

Procedures

These procedures were written for the Fisher Mini-Submarine gels. They may need to be slightly modified for other equipment.

Preparing Agarose

1. Prepare a mixture of 0.8% w/v agarose in TRIS-glycine pH 8.6 buffer. Make 40 mL of the solution in a flask.

2. Heat the flask in a microwave oven until the agarose has melted. Swirl to mix. This takes 30–40 s on high power in most microwave ovens for a single flask. Do not let the agarose boil too long, or it will boil over.

3. Allow the solution to cool for 2 min before pouring into the casting trays. It should be clear with no significant agarose clumps.

Pouring Agarose

1. Insert the rubber stoppers into the casting tray with the thick sides up.

2. Pour 40 mL of agarose into the casting tray.

3. Insert the comb into the gel. Make sure that the orientation is correct. The comb should not rest against the rubber stopper but should be about 1 cm from it.

4. Let the gel cool for 15 min. Ever so carefully, remove the rubber stoppers before removing the comb. You may have to use a spatula to break the surface tension between the agarose and the rubber.

Preparing and Applying the Sample

1. For each LDH sample to be loaded, take 10 μL of sample and add 2 μL of 6× sample buffer. Use the vortexes to mix the contents of the tube. Spin the tubes in a microfuge for a couple of seconds to bring the entire sample back down. The samples should be loaded as follows:

   Lane 1: LDH 1 isozyme

   Lane 2: LDH 2 isozyme

   Lane 3: LDH 5 isozyme

   Lane 4: Crude bovine heart LDH

   Lane 5: LDH isotrol (LDH 1–5)

   Lane 6: Your best sample

   Lane 7: Another of your samples

   Lane 8: Another of your samples

2. Transfer the casting trays to the electrophoresis unit with the sample wells nearest the black (−) electrode.

3. Slowly fill the chamber with electrophoresis buffer until the buffer is about 1 cm over the top of the gel.

4. Using a Pipetman, apply the samples to the wells. Load the entire sample if it will fit without overflowing but do *not* let the solution overflow into the other wells.

5. Put on the lid and connect the leads to the power supply. Electrophorese at 100 V until the marker bromophenol blue dye is close to running off the end. At 125 V, the separation takes about 1¼ h. However, the gel gets very hot, and the bands do not give as clean a separation. At 80 V, the separation takes 2–3 h, but the results are better.

Staining

1. Obtain a small plastic box with a lid. Carefully slide the gel out of the casting tray and into the box.

2. Cover the gel with about 25 mL of LDH activity stain.

3. Incubate the gel for 5–15 minutes in a 37°C water bath until the bands develop. Remove from the stain when fully developed.

4. Photograph or sketch the gel.

Name _____   Section _____

Lab partner(s) _____   Date _____

Analysis of Results and Questions

Experiment 9a: **Native Gel Separation of LDH Isozymes (Comprehensive Version)**

Data

Describe the results of your separation by including a sketch or photo of the gel. Be sure to correctly label each lane with the sample that was applied there. Sketch it so that the wells are at the top.

Analysis of Results

1. Why do some lanes have more than one band in them? What does each band represent?

2. Describe the nature of the LDH molecules that are separated on this system. How do you know that only one variable is at work here?

3. Which of your samples is the most active? Did you get the isozyme pattern you expected? Why or why not?

Questions

1. Describe the effects the following have on the way your gel is run. Use the following equation to justify your answer:

$$v = \frac{qE}{f}$$

   a. Increasing the agarose concentration to 1.5%

   b. Increasing the pH of the TRIS-glycine buffer to 9.5

   c. Decreasing the voltage at which the gel was run

2. You need to load 10 μg of protein into one of the wells of a gel. This needs to be in 1× buffer and in a total volume of 15 μL. You are given a 10 μg/μL solution of protein, a 5× buffer, and water. How much of each should you mix to load the gel correctly?

Experiment 9b

SDS-PAGE (Short Version)

In this experiment, you will use SDS-PAGE to separate proteins by molecular weight.

Prelab Questions

1. What molecular weight will you find for the LDH band?

2. What band will it be closest to of your standards?

3. Describe the procedure to use after the stacking gel has hardened and before you load your samples.

Objectives

Upon successful completion of this lab, you will be able to

- Pour polyacrylamide gels and assemble the gel chamber.
- Prepare and load protein samples into the sample wells.
- Stain and destain the gels with Coomassie Blue.
- Determine the molecular weights of unknown proteins.

Experimental Procedures

Materials

Gel boxes, plates, combs, spacers

Separating gel and stacking gel solutions

Reservoir buffer

BSA, 1 mg/mL (MW 66,000)

Ovalbumin, 1 mg/mL (MW 45,000)

(MW 36,000) Glyceraldehyde-3-phosphate dehydrogenase, 1 mg/mL

Carbonic anhydrase, 1 mg/mL (MW 29,000)

Trypsinogen, 1 mg/mL (MW 24,000)

Dalton Mark VII Standard Proteins mixture [1 mg/mL each BSA, ovalbumin, glyceraldehyde-3-phosphate dehydrogenase, carbonic anhydrase, trypsinogen, trypsin inhibitor (MW 20,000) and lactalbumin (MW 14,000)]

Unknown proteins

Methods

Preparing the Gel

These procedures are for the Bio Rad Mini Protean II. Your procedures may need to be revised to accommodate a different setup.

1. Set up the gel plates per Bio Rad instructions. Do this part very carefully. If you do not get the plates lined up properly or do not use the plate-leveling part of the casting tray, the gel will leak when poured.

2. After leveling the plates and setting the plate assembly in the casting tray, insert the comb and mark the plate on the outside 2 mm below the bottom of the comb. The bridge of the comb should come to rest on top of the gray plate spacers.

3. Prepare 1 mL of 10% w/v ammonium persulfate (AP) in water and store on ice. Note that 1 mL is enough for the entire class.

4. Use a Pipetman to add 17 μL of the AP to 5 mL of separating gel. Swirl gently to mix and quickly inject the solution between the plates up to the level of the mark that you made. This would be a bad time to discover that your pipet of choice does not fit into the vessel containing the solidifying acrylamide. A liquid transfer syringe or 5-mL pipet with pipet pump works well here.

5. If no leaks occur, use a Pasteur pipet to layer a few millimeters of butanol on top of the gel.

6. When the gel has hardened, remove the butanol and wash with reservoir buffer.

7. Add 13 µL of AP to 2.5 mL of the stacking gel and swirl gently.

8. Inject the stacking gel to the top of the shorter glass plate and insert the comb to the point that you marked before. Try to avoid having air bubbles stick to the comb. Draw on the outside of the plate to mark where the wells are.

9. When the stacking gel has hardened (15 min), remove the comb, and wash the wells with reservoir buffer. Place the glass plate assembly into the central clamp assembly. Fill the upper (central) reservoir as full as possible and check for leaks. Fill the tank with reservoir buffer up to the level of the lower electrode. The gel is now ready for sample application.

Preparing and Applying the Sample

1. Boil each protein standard or unknown protein listed in step 3 for 5 min and then place on ice. Spin the tubes so that the entire sample is at the bottom. These samples have already had the sample buffer/tracking dye added. Boiling denatures the proteins and allows the SDS and β-mercaptoethanol access to the whole protein.

2. Load 10 µL of the samples into the wells. Load slowly, allowing the solution to layer on the bottom of the well. *This is your last chance to avoid making the common error of loading samples before filling the reservoir with buffer.*

3. Load the samples in this order:

   Lane 1: Sample buffer only

   Lane 2: Dalton VII mixed markers (contains all markers)

   Lane 3: BSA, MW 66,000

   Lane 4: Ovalbumin, MW 45,000

   Lane 5: Glyceraldehyde-3-phosphate dehydrogenase, MW 36,000

   Lane 6: Carbonic anhydrase, MW 29,000

   Lane 7: Trypsinogen, MW 24,000

   Lane 8: Unknown protein

   Lane 9: LDH

   Lane 10: Dalton VII mixed markers

   If you have problems with one of the lanes, use the outside lane to repeat that sample.

TIP 9.6   Before staining your gel, make sure that you can identify which side is left and which is the top. Once it has rolled around in Coomassie Blue for a while, it might not be so easy.

4. Now you should be ready for electrophoresis. Before beginning, however, check to ensure that you don't have a slow leak from the upper reservoir chamber to the lower. If the upper chamber now has less buffer in it, you have to add more buffer to the top. One way to minimize the effects of leaking is to add more buffer to the lower chamber until the levels are almost equal between the height of the lower buffer and the upper buffer.

Running the Electrophoresis

1. Put the top on the unit, connecting the red leads to the red plugs on the power supply and the black to the black.

2. Have a teaching assistant or the instructor check the apparatus.

3. Electrophorese at 200 V until the dye front reaches the bottom (30–45 min).

4. Label a plastic container for staining and destaining.

Staining and Destaining

1. Separate the plates and loosen the gel from the plate by squirting water under it with a liquid-transfer syringe. Place the gel into the plastic container and overlay with Coomassie Blue protein stain.

2. The gel will be stained overnight and then placed into destain.

3. To view the gel, place the gel on the lid, not on a paper towel.

4. Sketch or photograph the gel.

Name _____

Section _____

Lab partner(s) _____

Date _____

Analysis of Results

Experiment 9b: **SDS-PAGE (Short Version)**

Data

Unknown Protein

1. Make a sketch or take a picture of your gel, labeling the protein lanes.

2. Calculate the R_m's for the bands. This is best done using the Dalton VII lane. The single standard lanes are used to verify the identity of a band in the Dalton mixture.

Protein	R_m
BSA	_____
Ovalbumin	_____
Glyceraldehyde-3-phosphate dehydrogenase	_____
Carbonic anhydrase	_____
Trypsinogen	_____
Unknown	_____
LDH	_____

Analysis of Results

1. Plot log MW versus Rm for the bands and turn in the graph with this report.

2. Determine the molecular weight of the unknown.

Questions

1. The last gel you did separated proteins based on charge. What is the variable this time? How do you know only one variable is at work here?

2. You have an enzyme that is composed of three subunits. Two subunits are 28 kD, and the other is 14 kD. How many bands do you see on SDS-PAGE?

3. Describe the results if you do an experiment to determine the physical properties of tyrosinase. If you run gel filtration and SDS-PAGE, what results do you see?

4. What aspects of your experiment are controlled by TEMED and ammonium persulfate?

5. What aspects of your experiment are controlled by acrylamide concentration?

Experiment 9c

SDS-PAGE (Comprehensive Version)

In this experiment, you will use SDS-PAGE to separate proteins by molecular weight and to verify the purity of your samples.

Objectives

Upon successful completion of this lab, you will be able to

- Pour polyacrylamide gels and assemble the gel chamber.
- Prepare and load protein samples into the sample wells.
- Stain and destain the gels with Coomassie Blue.
- Determine the molecular weights of unknown proteins.
- Verify the purity of your best LDH samples.

Experimental Procedures

Materials

Acrylamide/bisacrylamide,* 40%

TEMED[†]

4× separating buffer

4× stacking buffer

Reservoir buffer

2× sample buffer[†]

Dalton Mark VII protein mixture [1 mg/mL each BSA, ovalbumin, glyceraldehyde-3-phosphate dehydrogenase, carbonic anhydrase (MW 29,000), trypsinogen (MW 24,000), trypsin inhibitor (MW 20,000), and lactalbumin (MW 14,000)]

BSA, 1 mg/mL (MW 66,000)

Ovalbumin, 1 mg/mL (MW 45,000)

Glyceraldehyde-3-phosphate dehydrogenase, 1 mg/mL (MW 36,000)

---

* Acrylamide is a neurotoxin. You must wear gloves at all times when handling it or any solution containing it.
† TEMED and the sample buffer will be used in the hood.

Bio Rad Mini Protean II electrophoresis apparatuses

Power supplies

Procedures

These procedures were written for the Bio Rad Mini Protean II electrophoresis unit. You may need to modify them for different equipment.

Preparing the Gel

1. Prepare 10 mL of separating gel and 5 mL of stacking gel by mixing the ingredients according to the following table. Do *not* add the ammonium persulfate yet. Store the gel solutions on ice. Note that this is enough for two gels.

Ingredients for Gels

Reagent	Volume (12% separating gel)	Volume (5% stacking gel)
4× separating buffer	2.5 mL	None
4× stacking buffer	None	1.25 mL
H$_2$O	4.5 mL	3.1 mL
Acyrlamide/bisacrylamide	3.0 mL	0.65 mL
TEMED	6.7 μL	7.5 μL
10% ammonium persulfate	33 μL	25 μL

2. Clean the plates with great obsessiveness. Set up the gel per Bio Rad instructions. The initial set up in the casting tray will determine whether your gel leaks.

3. Prepare 1 mL of 10% ammonium persulfate (AP) and store on ice. Note that this volume is sufficient for the entire class.

4. Place the comb into the gel plates and insert until the outermost stop comes to rest on top of the gray spacers. With a fine-point permanent-ink pen, mark the glass plate at the bottom of the comb.

5. Divide the separating gel into two 5-mL aliquots and add 17 μL of AP to 5 mL of the separating gel. Swirl gently but thoroughly and quickly inject the solution between the plates up to the level of the mark that you made.

6. If no leaks occur, use a Pasteur pipet to layer a few millimeters of butanol onto the gel. Wait until the gel has hardened (15 min).

7. Remove the buffer and butanol that is on top of the separating gel and wash several times with an electrode buffer.

8. Divide the stacking gel into two 2.5-mL aliquots and add 13 μL of AP to one of them. Vortex gently.

9. Inject the stacking gel to the top of the shorter plate and insert the comb. Try to avoid having air bubbles stick to the comb.

10. When the stacking gel has hardened (15 min), remove the comb and then remove the plate assembly from the casting tray and attach the plate assembly to the central block. Place the central unit in the tank and fill the upper reservoir with reservoir buffer. Check for leaks. The upper reservoir should be filled to the top of the outer plates. The lower reservoir must be filled to the height of the lower electrode wire, about 1 cm from the bottom of the gel. The gel is now ready for sample application. If buffer leaks from the upper reservoir into the lower, the problem can be diminished by raising the level of the buffer in the lower reservoir.

Preparing and Applying the Sample

1. For each unknown sample, mix 10 μL with 10 μL of 2× sample buffer in a microcentrifuge tube. The standards have already been mixed with the 2× sample buffer, so pipet just 20 μL into the microcentrifuge tube (40 μL for the Dalton mixture).

2. Boil for 2 min and then place on ice. Spin the tubes so that the entire sample is at the bottom.

3. Load 15 μL of the samples into the wells. Load slowly, allowing the solution to layer on the bottom of the well.

4. Load the samples in the following order:

   Lane 1: Dalton mixed markers

   Lane 2: BSA

   Lane 3: Ovalbumin

   Lane 4: Glyceraldehyde-3-phosphate dehydrogenase

   Lane 5: Your 20,000 × g supernatant

   Lane 6: Your 65% ammonium sulfate pellet

   Lane 7: Your pooled Q-Sepharose fraction

   Lane 8: Your pooled Cibacron Blue fraction

   Lane 9: Your pooled gel filtration fraction

   Lane 10: Dalton mixed markers

Running the Electrophoresis

1. Put the top on the unit, connecting the red leads to the red plugs on the power supply and the black to the black.

2. Have a teaching assistant or the instructor check the apparatus.

3. Electrophorese at 180 V until the dye front reaches the bottom (45 min). Using a lower voltage will improve the results as it did with the agarose gels.

Staining and Destaining

1. Label (with tape) a plastic container for staining and destaining.

2. Remove the gel from the plates by injecting water with a liquid-transfer pipet between the gel and plate.

3. Place the gel in the tray with staining solution.

4. The teaching staff may put the gel in the destain for you, or you may come back and do it yourself. You need to come in later to observe the results.

Name ______________________ Section ______________________

Lab partner(s) ______________________ ______________________ Date ______________________

Analysis of Results

Experiment 9c: SDS-PAGE (Comprehensive Version)

Data

1. Make a sketch or take a picture of your gel, labeling the protein lanes.
2. Calculate the R_m's for the bands. This is best done using the Dalton VII lane. The single standard lanes are used to verify the identity of a band in the Dalton mixture.

Protein	R_m
BSA	________
Ovalbumin	________
Glyceraldehyde-3-phosphate dehydrogenase	________
Carbonic anhydrase	________
Trypsinogen	________
LDH	________

Analysis of Results

1. Plot log MW versus Rm for the bands and turn in the graph with this report.
2. Determine the molecular weight of the LDH monomer.

3. Is the molecular weight of the LDH monomer what you expected? Why or why not?

4. Which of your samples is the most pure? Is this what you expected? Why or why not?

5. What aspects of your experiment are controlled by TEMED and ammonium persulfate?

6. What aspects of your experiment are controlled by the concentration of acrylamide?

Additional Problem Set

1. Explain the purpose of the components of an acrylamide gel.

2. Which components of an acrylamide gel affect the qualities of the final product, and which components affect the speed of polymerization?

3. If you run Dalton VII mixed markers on an acrylamide gel and only five bands show up, what are some explanations for what happens to the other two?

4. What is the purpose of running both Dalton VII mixed markers and lanes with individual markers on SDS-PAGE?

5. Given the results that you see with the Dalton markers, is it possible to use a 12% separating gel to run a set of proteins that range in molecular weight from 10,000 to 200,000? Explain.

6. What causes a protein to migrate on SDS-PAGE in such a way that it appears to be a protein that is much bigger?

Webconnections

For a list of websites related to the material covered in this chapter, go to **Webconnections** at the *Experiments in Biochemistry* site on the Brooks/Cole Publishing website. You can access this page at http://www.brookscole.com and follow the links from the chemistry page.

References and Further Reading

Acquaah, G. *Practical Protein Electrophoresis for Genetic Research*. Portland: Dioscorides Press, 1992.

Allen, R. C. *Gel Electrophoresis of Proteins and Nucleic Acids: Selected Techniques*. Berlin: W. de Gruyter, 1994.

Anderson, J. N. *A Laboratory Course in Molecular Biology*. Vol. 1, *Physiology and Clinical Biology of Proteins*. Dayton, IN: Modern Biology, 1986.

Bollag, D. M., M. D. Rozycki, and S. J. Edelstein. *Protein Methods*. New York: Wiley-Liss, 1996.

Boyer, R. F. *Modern Experimental Biochemistry*. Menlo Park, CA: Addison-Wesley, 1993.

Cahn, R. D., N. O. Kaplan, L. Levine, and E. Zwilling. "Nature and Development of Lactic Dehydrogenases." *Science* 136 (1962).

Camilleri, P. *Capillary Electrophoresis: Theory and Practice*. Boca Raton, FL: CRC Press, 1993.

Campbell, M., and S. Farrell. *Biochemistry*. Belmont, CA: Brooks/Cole, 2005.

Clausen, J., and B. Ovlisen. "Lactate Dehydrogenase Isoenzymes of Human Semen." *Biochemical Journal* 97 (1965).

Dickerson, R. E., and I. Geiss. *The Structure and Action of Proteins*. New York: Harper and Row, 1969.

Dryer, R. L., and G. F. Lata. *Experimental Biochemistry*. New York: Oxford University Press, 1989.

Foret, F., L. Krivankova, and P. Bocek. *Capillary Zone Electrophoresis.* New York: VCH Publishers, 1993.

Gersten, D. M. *Gel Electrophoresis: Proteins—Essential Techniques.* New York: Wiley, 1996.

Hawcroft, D. M. *Electrophoresis.* Oxford: IRL Press/Oxford University Press, 1997.

Jones, P. *Gel Electrophoresis: Nucleic Acids.* New York: Wiley, 1995.

Kaufman, P. B. *Handbook of Molecular and Cellular Methods in Biology and Medicine.* Boca Raton, FL: CRC Press, 1995.

Manchenko, G. P. *Handbook of Detection of Enzymes on Electrophoretic Gels.* Boca Raton, FL: CRC Press, 1994.

Martin, R. *Gel Electrophoresis.* Oxford: Oxford Scientific, 1996.

Patel, D. *Gel Electrophoresis.* New York: Wiley, 1994.

Rabilloud, T. "Silver Staining of Proteins in Polyacrylamide Gels: A General Overview." *Cellular and Molecular Biology* 40 (1994).

Robyt, J. F., and B. J. White. *Biochemical Techniques.* Long Grove, IL: Waveland Press, 1990.

Schoeff, L. E., and R. H. Williams. *Principles of Laboratory Instruments.* St. Louis: Mosby Year Book, 1993.

Weber, K., and M. Osborn. "The Reliability of Molecular Weight Determinations by SDS-PAGE." *Journal of Biological Chemistry* 244, no. 16 (1969).

Westermeier, R. *Electrophoresis in Practice: A Guide to Theory and Practice.* New York: VCH Publishers, 1993.

Whitmore, D. H. *Electrophoretic and Isoelectric Focusing Techniques in Fisheries Management.* Boca Raton: CRC Press, 1990.

Chapter 10

Western Blots

TOPICS

Introduction

---

*In this chapter, we discuss the popular technique of Western blotting. A **Western blot** is a transfer of proteins from an electrophoresis gel (usually polyacrylamide) onto a thin membrane of nitrocellulose or some other absorbent material. Antibodies are then used to locate desired proteins so that their position on the original gel can be determined.*

10.1 Western Blot Theory

In the native gel electrophoresis experiment, we separated LDH isozymes on a gel and then identified their locations via an activity stain. That was one way to pick a desired protein out of a mixture because many of the samples that we separated were crude and contained other proteins. What do you do if the desired protein is not an enzyme? There is no activity stain that you can use. The answer is to use antibodies and the process called Western blotting.

A Western blot involves running a regular protein-separating gel, either native or SDS. Once the proteins are separated, make a blot with the gel and a membrane such as nitrocellulose that binds proteins. Then put the gel–membrane blot into a different type of electrophoresis apparatus that will transfer the proteins out of the gel and onto the membrane. Once this is done, use specific antibodies to find the protein of interest on the membrane. Figure 10.1 shows the gel–membrane sandwich. Figure 10.2 shows typical results of transferring proteins to a membrane.

Whether you use a native gel or an SDS gel depends on the antibodies you will use later. Some antibodies react to an epitope that is present only in the native conformation. Other antibodies will react to a protein even if it is completely denatured by SDS. This happens when the antibody reacts to a small linear amino acid sequence.

FIGURE 10.1 *Western blot sandwich*

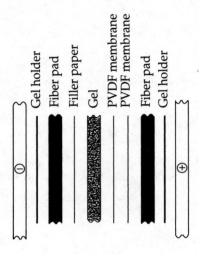

10.2 Antibodies

The key to the Western blot technique lies in the specificity of the antibodies chosen. Once the proteins are transferred onto the nitrocellulose, they are reacted with two different types of antibodies, as shown in Figure 10.3. The first reaction is with a primary antibody. This is an antibody against the protein that we are looking for. It may be a monoclonal antibody, which is very pure and expensive. These are derived from a single B cell, which produces only one type of specific antibody.

Polyclonal antibodies are a mixture collected from serum and are not as specific or expensive. This is shown in Figure 10.3 as step 2. Wherever the primary antibody finds the protein of interest on the membrane, it will bind. Unfortunately, we cannot see those results, so we have to do another step. The next step is to react the blots with a secondary antibody.

FIGURE 10.2 *Coomassie Blue total protein–stained gel (left) and Western blot (right)*
(Courtesy of Bio Rad)

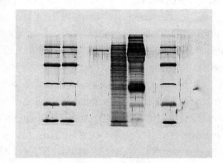 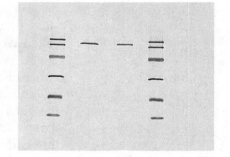

TIP 10.1 Be very careful that you add the antibodies in the correct order. In fact, do not even pick up the secondary antibody until your incubation with the primary antibody is over. Also, many antibodies are labeled very similarly. For example, if you are to use goat antimouse IgG-HRP conjugate for a secondary antibody, the experiment will not work if you accidentally pick goat antimouse IgG-AP conjugate instead.

FIGURE 10.3 *Antibody reactions in Western blots*

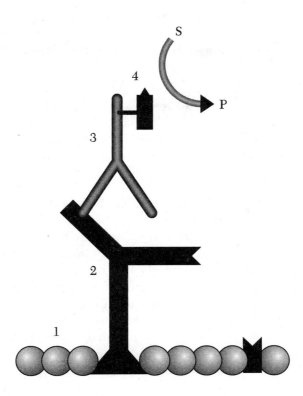

The secondary antibody is an antibody against the general class of primary antibodies. For instance, if we use a mouse primary antibody against human LDH, then our secondary antibody could be a goat antibody against mouse antibodies. The secondary antibody will bind to the blot where there is bound primary antibody. This is indicated in Figure 10.3 as step 3.

10.3 Color Development

The protein–primary antibody–secondary antibody complex is not visible either. However, most secondary antibodies are conjugated to an enzyme or another available marker. Common enzymes are alkaline phosphatase or horseradish peroxidase. This is shown as step 4 in Figure 10.3. These enzymes catalyze reactions that leave visible products. We eventually incubate the membrane in the substrates for the conjugated enzymes and get a visible band. Horseradish peroxidase is the first label used with secondary antibodies. HRP-labeled antibodies catalyze the following reaction:

$$\text{4-chloro-1-naphthol}_{red} + H_2O_2 \rightarrow \text{4-chloro-1-naphthol}_{ox} + H_2O + \frac{1}{2}O_2$$

where the oxidized form of naphthol forms a purple precipitate. This secondary antibody gives very clean results with little background staining, but the bands are very light and often difficult to see.

Alkaline phosphatase is another common enzyme tag used with secondary antibodies. It catalyzes the following reaction:

5-bromo-4-chloro-3-indolyl phosphate + nitroblue tetrazolium$_{ox}$ →

5-bromo-4-chloro-3-indole + P_i + nitroblue tetrazolium$_{red}$

where the reduced form of NBT forms a purple precipitate, as it did with the LDH activity stain we used in Experiment 9a.

10.4 Blocking and Washing

Interspersed between the antibody reactions are some other steps. If we transfer the proteins onto the nitrocellulose and they immediately react with the primary antibodies, they bind all over the nitrocellulose because this support has a high affinity for all proteins. Blocking is a procedure we use after we do the transfer. Blocking entails suspending the membrane in a solution of nonspecific proteins that will then bind up all sites on the nitrocellulose that have not already been bound by the transferred proteins. Common blocking reagents are milk, BSA, or gelatin.

After each incubation with an antibody, an extensive washing procedure rids all antibodies that are not bound tightly to the transferred proteins. Without such washing, extensive background color will be seen when the color development solution is used, and seeing the bands will be difficult.

10.5 Why Is This Important?

The importance of molecular biology and immunology are made more apparent every day as scientists fight the AIDS virus. Techniques that allow the study of DNA and the proteins produced from it grow in importance. Several blotting techniques are used routinely. Southern blots are used to transfer DNA, Northern blots for RNA, and Western blots for proteins. All three techniques are similar, and all are based on separating molecules on a gel and then transferring them onto nitrocellulose. Specific molecules can then be identified by reacting the blots with a probe. For Northern or Southern blots, this probe is a sequence of nucleotides. For Western blots, the probe is an antibody specific for the protein of interest. Once the location of the protein is established, it can be excised from the gel and studied further.

ESSENTIAL INFORMATION

Western blot is a technique for identifying and quantifying a particular protein out of a mixture. Typically, an acrylamide gel is used to separate proteins according to molecular weight or charge via SDS-PAGE or native gel electrophoresis. Then the gel is put into a sandwich, and a second electrophoresis drives proteins out of the gel and onto a membrane such as nitrocellulose.

Once proteins are isolated and concentrated on the membrane, very specific antibodies are used to probe for the presence of the protein of interest. A very specific primary antibody binds to the protein. A secondary antibody then binds to the primary antibody. This secondary antibody carries an enzyme or fluorescent marker that allows the protein–antibody complex to be seen.

Experiment 10

Western Blot of Serum Proteins

In this experiment, you will use native gel electrophoresis on polyacrylamide to separate serum proteins from several species. The separated proteins will be transferred to nitrocellulose via the Western blot procedure. The nitrocellulose will then be incubated with antibodies to locate the position of human serum albumin.

Prelab Questions

1. Briefly diagram the order of events for this experiment.

2. Indicate how you will mark the gels and blots so that you know which bands pertain to which sera.

Objectives

Upon successful completion of this lab, you will be able to

- Set up a Western blot apparatus and gel sandwich.
- React blots with primary and secondary antibodies.
- Perform color development reactions.
- Identify desired proteins by comparing blots to acrylamide gels.

Experimental Procedures

Materials

Acrylamide gels

2× sample buffer

Serum samples

Power supplies

Transfer buffer

Nitrocellulose

Coomassie Blue R protein stain

TBST (0.05% Tween 20 in TBS)

Milk solution (10% milk powder in TBST)

Mouse anti-HSA (human serum albumin)

Goat antimouse IgG

37°C water baths

Color developing solution

Western blot electrophoresis apparatuses

Methods

Lab Period 1

1. Acquire an electrophoresis apparatus with prepoured gel.

2. Mix each sample with an equal volume of 2× sample buffer (SDS free). Load each lane with the serum samples, about 10 μL per lane. Load the ten lanes of the gel in this order:

   Lanes 1, 6:   BSA

   Lanes 2, 7:   Human serum

   Lanes 3, 8:   Pig serum

   Lanes 4, 9:   Cow serum

   Lanes 5, 10: Horse serum

3. Electrophorese at 200 V until the dye front is close to the bottom of the gel.

4. Cut the gel in half lengthwise so that you have two gels with five lanes each. Make a notch in one of the corners so that you know the proper orientation.

5. Place one half into a tray with Coomassie Blue R protein stain.

6. Wet all materials for the gel sandwich with transfer buffer. *Wear gloves during all handling of the gel, blotting papers, and nitrocellulose.*

7. Build the gel sandwich with blotting paper and nitrocellulose with two pieces of filter paper, the gel, one piece of nitrocellulose, and two more pieces of filter paper. All should be the same size as the gel. Put a notch in the nitrocellulose so that you know its orientation. Place the sandwich in the transfer apparatus, making sure the nitrocellulose is toward the positive electrode.

8. Prepare a plastic tray with 30 mL of blocking solution. Label the tray with your name.

9. At some point the next day, the Coomassie Blue–stained gel must be put into a destain solution (10% acetic acid). This will be used later to see all the proteins.

Lab Period 2

1. Wash the nitrocellulose membrane a few times with distilled water and then with TBST. Pour off the last washing of TBST.

2. Place the nitrocellulose membrane into a tray with 20 mL of primary antibody (mouse anti-HSA). Incubate 20 min on a shaker.

3. Wash the membrane three times for 10 min each in 20 mL of TBST to remove unbound antibody.

4. Replace the TBST with 20 mL of secondary antibody (goat antimouse IgG-HRP conjugate).

5. Incubate for 20 min as before.

6. Pour off the secondary antibody and wash three times for 10 min each with 20 mL of TBST.

7. Blot the membrane damp dry on filter paper and transfer it to 20 mL of the color development solution. Reactive areas will turn purple in about 15 min but will continue for up to 4 h.

8. When the blot has developed to a desired intensity, stop the reaction by rinsing the membrane with distilled water. Air dry the membrane on filter paper for storage and protect it from light. The band will fade over time.

9. Compare the nitrocellulose to the Coomassie Blue–stained gel.

Name _____ Section _____

Lab partner(s) _____ Date _____

Analysis of Results

Experiment 10: **Western Blot of Serum Proteins**

Data

Draw a picture of your gel and blot, showing the location of the bands and the identification of the lanes. Use a separate sheet of paper or turn in a photograph if you have one.

Calculations and Questions

1. Which of the bands on the native gel represents serum albumin?

2. From this experiment, what is the similarity and difference between serum albumin from the different species tested?

3. Suppose, upon developing your blot, you see more than one band. What are some possible reasons for this?

4. Suppose, upon developing your blot, you see no bands at all. What are some possible reasons for this?

5. How can you determine which of the possibilities that you suggested in Question 4 is the correct one?

6. How can you test the efficiency of transfer of the proteins from the gel to the nitrocellulose membrane?

7. Give the mechanism or purpose of the following steps in the blotting procedure. Why did you do it, and/or how does it work?

   a. Tagging the secondary antibody with horseradish peroxidase

   b. Soaking in milk solution

   c. Running a native acrylamide gel

8. Why must the secondary antibody be used? Wouldn't it be simpler to just put the HRP conjugate directly on the primary antibody?

Experiment 10a

Western Blot of LDH

In this experiment, you will separate your LDH samples on a native polyacrylamide gel, transfer them to nitrocellulose via the Western blot procedure, and then probe for LDH using an antihuman LDH-H subunit antibody.

Objectives

Upon successful completion of this lab, you will be able to

- Set up a Western blot apparatus and gel sandwich.
- React blots with primary and secondary antibodies.
- Perform color development reactions.
- Identify desired proteins by comparing blots to acrylamide gels.

Experimental Procedures

Separation of LDH Isozymes (Day 1)

Materials

Acrylamide/bisacrylamide,* 40%

TEMED[†]

$4\times$ separating buffer (SDS free)

$4\times$ stacking buffer (SDS free)

TRIS-glycine reservoir buffer (SDS free)

$2\times$ sample buffer (SDS free)[†]

Bio Rad Mini Protean II electrophoresis apparatuses

Power supplies

Human LDH sample

Bovine LDH sample

Human serum

Your LDH samples

* Acrylamide is a neurotoxin. Wear gloves at all times when handling it or any solution containing it.
† TEMED and the sample buffer will be used in the hood.

TRIS-glycine, MeOH transfer buffer

TRIS buffered saline (TBS)

TRIS buffered saline + Tween 20 (TBST)

Western blot electrophoresis apparatuses

Procedure

1. Prepare a 10% acrylamide gel according to the following table.

Reagent*	Volume (10% separating gel)	Volume (5% stacking gel)
4× separating buffer	2.5 mL	None
4× stacking buffer	None	1.25 mL
H_2O	5.0 mL	3.1 mL
Acrylamide/ bisacrylamide	2.5 mL	0.65 mL
TEMED	6.7 μL	7.5 μL
10% ammonium persulfate	33 μL	25 μL

* Note all reagents are free of SDS

2. Mix each sample with an equal volume of 2× sample buffer (SDS free). Load each lane with a 15-μL sample. Load the ten lanes of the gel so that the first five contain one lane each of the five different samples and the second five lanes also contain one lane each sample.

3. Electrophorese as in Experiment 9b until the dye front runs off the gel.

4. Cut the gel in half lengthwise so that you have two gels with five lanes each. Make a notch in one of the corners so that you know the proper orientation.

5. Place one half into a tray with Coomassie Blue R protein stain.

6. Wet all materials for the gel sandwich with transfer buffer (TRIS-glycine reservoir buffer with 20% methanol). *Wear gloves during all handling of the gel, blotting papers, and nitrocellulose.*

7. Build the gel sandwich with blotting paper and nitrocellulose so that there are two pieces of filter paper, the gel, one sheet of nitrocellulose, and two more pieces of filter paper. Put a corresponding notch in the nitrocellulose so that you know its orientation. Place the sandwich in the electroblotting apparatus so that the nitrocellulose is toward the

positive electrode. Place a labeled container of blocking solution (30 mL of 10% milk powder in TBST) near the apparatus.

8. Transfer the proteins onto the nitrocellulose, following the Western blot manufacturer's instructions, and put your blot into the tray with blocking solution. Also stain the gel that was used for the transfer with Coomassie Blue.

9. Incubate for at least 1 h or until the next lab period. For overnight incubations, store the blocking solution at 4°C.

10. At some point, put the Coomassie Blue–stained gels into destain solution (10% acetic acid).

Reaction with Antibodies (Day 2)

Materials

TRIS buffered saline (TBS)

TRIS buffered saline + Tween 20 (TBST)

Mouse monoclonal antihuman LDH-H subunit (diluted 1 : 5000 in TBST)

Goat antimouse IgG-AP conjugate (diluted 1 : 2500 in TBST)

Western blue AP substrate

Procedure

1. Wash the nitrocellulose membrane a few times with distilled water and then with TBST. Pour off the last washing of TBST.

2. Place the nitrocellulose membrane into a tray with 15 mL of diluted primary antibody in TBST. Incubate 30 min at room temperature on a shaker.

3. Wash the membrane three times for 10 min each in 20 mL of TBST to remove unbound antibody.

4. Replace the TBST with 15 mL of diluted secondary antibody.

5. Incubate at room temperature for 30 min as before.

6. Pour off the secondary antibody and wash three times for 10 min each with 20 mL of TBST.

7. Blot the membrane damp dry on filter paper and transfer it to 10 mL of the color development solution. Reactive areas will turn purple in 15–30 min.

8. When the blot has developed to a desired intensity, stop the reaction by rinsing the membrane with distilled water. Air dry the membrane on filter paper for storage and protect it from light.

Name _____   Section _____

Lab partner(s) _____   Date _____

Analysis of Results

Experiment 10a: **Western Blot of LDH**

Data

Draw a picture of your gel and blot, showing the location of the bands and the identification of the lanes. Use a separate sheet of paper or turn in a photograph if you have one.

Calculations and Questions

1. Which of the bands on the native gel represents LDH?

2. Which LDH isozymes show up on the blot?

3. What can you say about the specificity of the antibody to human LDH-H subunit?

4. Suppose, upon developing your blot, you see more than the bands that you expect. What are some possible reasons for this?

5. Suppose, upon developing your blot, you see no bands at all. What are some possible reasons for this?

6. What controls or other experiments can you run to determine what caused the problems described in Questions 4 and 5?

7. How can you test the efficiency of transfer of the proteins from the gel to the nitrocellulose membrane?

8. Give the mechanism or purpose of the following steps in the blotting procedure. Why did you do it, and/or how does it work?

   a. Tagging the secondary antibody with alkaline phosphatase

   b. Soaking in milk solution

   c. Running a native acrylamide gel

9. Why do most Western blot experiments use two antibodies? Why not just tag the primary antibody with the alkaline phosphatase?

Additional Problem Set

1. If you are trying to design a Western blot experiment, how do you determine the time and voltage needed to transfer proteins to the nitrocellulose membrane?

2. When a native acrylamide gel is used for a Western blot, the proteins that run the farthest down the gel usually transfer to nitrocellulose the fastest. Why?

3. Sometimes Western blots are done with native gels and other times with an SDS-PAGE. What influences your choice of gel type?

4. What is the difference between a monoclonal antibody and a polyclonal antibody? Why might you choose one or the other as part of a Western blot experiment?

5. Why is it beneficial to use a polyclonal antibody as a secondary, enzyme-labeled antibody for a Western blot experiment?

6. During the experiment, the gel used to transfer proteins to the nitrocellulose also had to be stained with Coomassie Blue. What is the purpose of this step?

Webconnections

For a list of websites related to the material covered in this chapter, go to **Webconnections** at the *Experiments in Biochemistry* site on the Brooks/Cole Publishing website. You can access this page at http://www. brookscole.com and follow the links from the chemistry page.

References and Further Reading

Bio Rad. *Life Science Research Products Catalog,* 2004.

Campbell, M., and S. Farrell. *Biochemistry.* Belmont, CA: Brooks/Cole, 2005.

Chattopadhya, D., R. K. Aggarwal, U. K. Baveja, V. Doda, and S. Kumari. "Evaluation of Epidemiological and Serological Predictors of Human Immunodeficiency Virus Type-1 Infection among High-Risk Professional Blood Donors with Western Blot Indeterminate Results." *Journal of Clinical Virology* 11, no. 1 (1998).

Cruz Sui, O., M. T. Perez Guevara, M. Izquierdo Marquez, L. Lobaina Batelemy, I. Ruibal Brunet, and E. Silva Cabrera. "Evaluation of a Western Blot System for the Confirmation of HIV-1 Antibodies." *Revista Cubana de Medicina Tropical* 49, no. 1 (1997).

Dartsch, C., and L. Persson. "Recombinant Expression of Rat Histidine Decarboxylase: Generation of Antibodies Useful for Western Blot Analysis." *International Journal of Biochemistry and Cell Biology* 30, no. 7 (1998).

Dunbar, B. S. *Protein Blotting: A practical Approach.* Great Britain: IRL Press/Oxford University Press, 1994.

Egger, D., and K. Bienz. "Colloidal Gold Staining and Immunoprobing on the Same Western Blot." *Methods in Molecular Biology* 80 (1998).

Foschini, M. P., S. Macchia, L. Losi, A. P. Dei Tos, G. Pasquinelli, L. di Tommaso, S. Del Duca, F. Roncaroli, and P. R. Dal Monte. "Identification of Mitochondria in Liver Biopsies. A Study by Immunohistochemistry, Immunogold and Western Blot Analysis." *Virchows Archives* 433, no. 3 (1998).

Frost, F. J., A. A. de la Cruz, D. M. Moss, M. Curry, and R. L. Calderon. "Comparisons of ELISA and Western Blot Assays for Detection of *Cryptosporidium* Antibody." *Epidemiology of Infections* 121, no. 1 (1998).

Jack, R. C. *Basic Biochemical Laboratory Procedures and Computing.* New York: Oxford University Press, 1995.

Reiche, E. M., M. Cavazzana, H. Okamura, E. C. Tagata, S. I. Jankevicius, and J. V. Jankevicius. "Evaluation of the Western Blot in the Confirmatory Serologic Diagnosis of Chagas' Disease." *American Journal of Tropical Medicine and Hygiene* 59, no. 5 (1998).

Robyt, J. F., and B. J. White. *Biochemical Techniques.* Long Grove, IL: Waveland Press, 1990.

Chapter 11

Restriction Enzymes

TOPICS

Introduction

In this chapter, we begin our study of molecular biology. All modern techniques of molecular biology and our study of DNA are made possible by the discovery of restriction enzymes (nucleases), enzymes that cut DNA at specific sequences. Hundreds of different restriction enzymes have now been isolated and purified from different bacteria.

11.1 Restriction Nucleases

Many types of enzymes modify nucleic acids. One particular type is a **nuclease,** and it acts to catalyze the hydrolysis of the phosphodiester backbone of nucleic acids. Some nucleases are specific for DNA and others for RNA. Some cut only from an existing end of the nucleic acid; these are called **exonucleases.** Others, **endonucleases,** cut from the inside. One specific type, **restriction endonucleases,** has played an important role in the development of modern recombinant DNA technology.

Much of what we know today about molecular biology is made possible by the discovery of restriction endonucleases from bacteria. Bacteria are often attacked by viruses (*bacteriophages*), which are dangerous to them. While studying bacteria and the phages that infect them, researchers discovered that bacteriophages that grew well in one type of bacteria often grew poorly in another. It was said that their growth was *restricted*. Researchers found out that differences between the DNA of the bacteria and that of the phage are based on certain nucleic acid bases that are methylated. The bacteria possess a type of enzyme that cleaves DNA at specific sites unless the DNA is methylated at those sites, as its own DNA is. Figure 11.1 shows this concept.

FIGURE 11.1 *Cleavage of foreign DNA and protection by methylation*

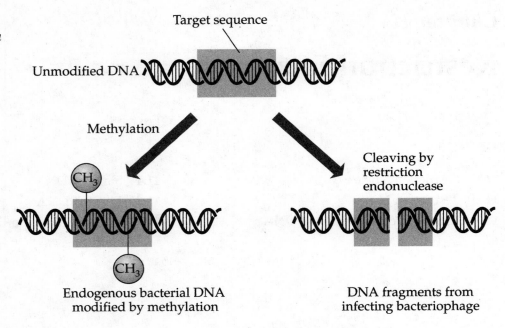

For example, the restriction enzyme from *Haemophilus influenzae* Rd, *Hind*III, cuts DNA at the following site:

$$\downarrow$$
5´-AAGCTT-3´
3´-TTCGAA-5´
$$\uparrow$$

whereas the restriction enzyme from *Escherichia coli* Ry13, *Eco*RI, cuts at the following site:

$$\downarrow$$
5´-GAATTC-3´
3´-CTTAAG-5´
$$\uparrow$$

Table 11.1 shows some typical restriction nucleases and where they cut. Note that some enzymes cut straight across from each other, leaving a **blunt end,** like *Hae*III; while most cut at offset positions, leaving **sticky ends.**

11.2 Restriction Maps

A restriction map is a drawing of a piece of DNA showing the location of the restriction sites for the enzymes of interest. DNA is numbered by virtue of its bases, with one end being arbitrarily called zero and the other end

TABLE 11.1 *Restriction Endonucleases and Their Cleavage Sites*

Enzyme*	Recognition and Cleavage Site
*Bam*HI	5′-G↓GATCC-3′ 3′-CCTAG↑G-5′
*Eco*RI	5′-G↓AATTC-3′ 3′-CTTAA↑G-5′
*Hae*III	5′-GG↓CC-3′ 3′-CC↑GG-5′
*Hind*III	5′-A↓AGCTT-3′ 3′-TTCGA↑A-5′
*Hpa*II	5′-C↓CGG-3′ 3′-GGC↑C-5′
*Not*I	5′-GC↓GGCCGC-3′ 3′-CGCCGG↑CG-5′
*Pst*I	5′-CTGCA↓G-3′ 3′-G↑ACGTC-5′

Arrows indicate the phosphodiester bonds cleaved by the restriction endonucleases.

* The name of the restriction endonuclease consists of a three-letter abbreviation of the bacterial genus and species from which it is derived—for example, *Eco* for *Escherichia coli.*

being the total number of bases in the piece of DNA. As an analogy, think of a ruler that has a division every millimeter for 300 mm. If we take a knife and cut the ruler at 50 mm and 150 mm, the knife is the equivalent of a restriction enzyme, and the 50- and 150-mm designations are the restriction sites. To make such a restriction map, we draw something like this:

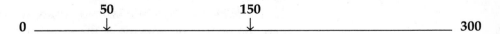

```
              50                  150
  0 ──────────↓──────────────────↓──────────────── 300
```

If we then cut the ruler with a pair of scissors at 100 mm, the scissors is the equivalent of a second restriction enzyme. If we want to make a restriction map of the two, it looks like this:

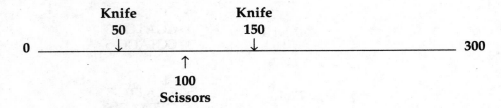

Restriction mapping is one of the most powerful and exciting techniques in molecular biology today. It is the basis of a related technique called **restriction fragment length polymorphism** (RFLP). Much has been learned about gene structure, especially mutated forms of genes, with RFLP technology. If necessary, the final step of paternity testing uses RFLPs. The cornerstone of modern forensic science, it is also used to prove or disprove the identity of a suspected criminal from blood or other tissue samples. Figure 11.2 shows the basics of RFLP.

11.3 Agarose Gel Separation of DNA

Agarose is the matrix of choice for separation of DNA fragments larger than 1000 base pairs. All DNA is of uniform shape and charge–mass ratio, so the fragments separate by molecular weight or size, which are measured as base pairs or kilobase pairs. Like molecular weight determination of proteins with SDS-PAGE, a plot of log MW of DNA versus distance migrated should give a straight line. *Should* doesn't always equate to *does*, however, so a curve may have to be drawn.

The fragments separate on the gel and are compared to a standard; that is the easy part. The difficult part is determining what the fragment pattern means. Here is where most students have problems with this type of experiment. Continuing with the ruler analogy, if we cut the ruler with the knife at position 50 mm and 150 mm, we get fragments that are 50 mm, 100 mm, and 150 mm long. If you cannot immediately see why this is so, do not read any further until you understand it.

If we can separate the ruler pieces on a gel, they will run in order with the smallest fragment (50 mm) running the fastest, followed by the middle piece (100 mm), and finally the largest piece (150 mm). When plotting fragment size versus distance migrated, we use the numbers 50, 100, and 150 because those are the lengths of the pieces. On the gel, we see three pieces. If we also run a gel of the pieces we obtained from cutting the ruler with the scissors, we see two pieces, one corresponding to 100 mm and one corresponding to 200 mm.

To make a complete restriction map, we often have to cut with more than one restriction enzyme at the same time. This is called a *double digest*.

FIGURE 11.2 *Restriction fragment length polymorphism of the β-globin gene (bp = base pair)*

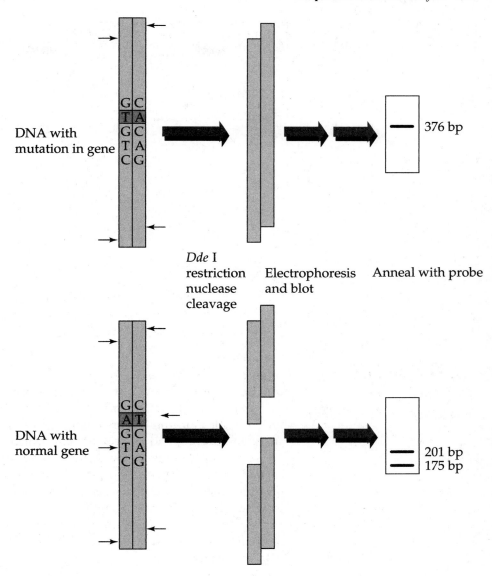

By cutting first with one enzyme, then another, and then both, we get valuable information about where the enzymes cut. In the ruler analogy, if we cut with both the scissors and the knife, we would see pieces of 50 mm and 150 mm. The 50-mm piece actually represents several different pieces of the same size, but they show up at the same place on the gel.

Figure 11.3 shows what the ruler analogy gel might look like if we did a traditional experiment designed to determine where the scissors and the knife cut. To do this type of experiment, we usually know where one of the enzymes cuts. It is the second one that we are trying to determine. We can do this by analyzing the gel from both single digests and the double digest.

FIGURE 11.3 *The ruler analogy for restriction digests*

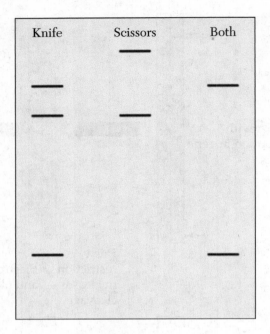

11.4 Staining DNA

The most common way to visualize DNA is by putting the chemical ethidium bromide (EtBr) into the gel or reservoir buffer. Ethidium bromide intercalates into DNA and fluoresces bright orange when exposed to UV light. If EtBr is in the buffer during the electrophoresis, then the progress can be monitored, and one can actually watch the bands separate.

The disadvantage to this is that *EtBr is a strong carcinogen and must be handled with extreme care.* For this reason, we often let EtBr soak into the gel after the electrophoresis is over so that EtBr is localized to one room. The downside is that this takes longer.

Because EtBr is toxic, companies such as Invitrogen have produced alternate DNA stains, such as SyBr Gold. These stains are as sensitive as EtBr but nontoxic. Gels stained with SyBr Gold can be viewed with a special filter and a UV lightbox.

The extent of the run can be monitored as usual by using tracking dyes. On a 1% agarose gel, bromophenol blue migrates with fragments of around 400 base pairs, whereas xylene cyanol migrates closer to 3000 base pairs.

11.5 Phage λ DNA

Bacteriophage λ, a virus that infects *E. coli*, is one of the most studied phages. The entire phage is 48,502 base pairs in length and contains restriction sites for many endonucleases. *Hind*III cuts λ DNA seven times at the positions shown in Table 11.2.

The fragments generated are based on the difference between two adjacent cuts or between a cut and either end. Also, the smallest fragment rarely shows up on a gel. The 564–base pair fragment usually does show up if the gel isn't run too long.

TABLE 11.2 Hin*dIII Digestion of* λ *DNA*

Cut Sites from Zero Point	Fragment Generated from Zero or from Last Cut (kb)
23,130	23.1
25,157	2.0
27,479	2.3
36,895	9.4
37,459	0.564
37,584	0.125
44,141	6.5
48,502 (end, not a site)	4.3

FIGURE 11.4 λ *DNA cut with* Hin*dIII*

A restriction map showing the *Hin*dIII sites on λ DNA looks like the following:

```
                                              37.6
                          25.1                37.4
              23.1        27.4                36.9            44.2
0 _____   23.1   ↓2   ↓2.3↓   9.4          ↓↓↓   6.5    ↓    4.3   48.5
```

When λ DNA is cut with *Hin*dIII, the resulting banding pattern looks like that shown in Figure 11.4. This is a very common set of DNA size markers. You can either create these markers yourself or buy commercially prepared λ DNA/*Hin*dIII fragments.

PRACTICE SESSION 11.1   If you cut λ DNA with *Hin*dIII and a second enzyme that cuts one time at position 33,498, what does the resulting gel look like?

Looking at the λ map shown above, we can see that the second enzyme cuts in-between two of the *Hin*dIII sites—namely, the ones that yield the 9.4-kb (kilobase) piece. The 9.4-kb piece is the second band on the gel shown in Figure 11.4. If we cut the λ DNA with just the second enzyme, it cleaves the DNA into two pieces. One would be 33,498 base pairs long, a big piece that is high up on the gel. The second piece is 48,502 − 33,498 = 15,004 base pairs.

This shows up lower than the first band from the *Hin*dIII cut (the 23.1 kb) but higher than the 9.4-kb piece. If we run a double digest, the majority of the bands from the *Hin*dIII digestion will be untouched because the second enzyme does not cleave in their regions. The only band that changes is the 9.4-kb fragment. It is cut by the second enzyme and does not show up in the double-digest lane. In place of the 9.4-kb fragment, two

ESSENTIAL INFORMATION

Restriction enzymes are used to cut DNA at very specific sequences. They are the backbone of molecular biology. When DNA is cut with restriction enzymes, fragments result that can be separated on an agarose gel. Using an agarose gel, the size of the fragments can be measured if standards are available to compare them to. A common DNA standard is generated by cutting phage λ DNA with *Hind*III. By cutting DNA with a second enzyme and comparing it to the first, we can determine where in the DNA the second enzyme cuts. Eventually, we can make a restriction map, which shows the exact location of restriction enzyme cut sites in a piece of DNA.

fragments are generated. The first is 33,498 − 27,479 = 6019 base pairs. The second is 36,895 − 33,498 = 3397 base pairs. Figure 11.5 shows what such a gel looks like. ●

11.6 Why Is This Important?

FIGURE 11.5 *λ DNA cut with* Hind*III (lanes 1 and 4, from left to right), a second enzyme (lane 3), and both* Hind*III and the second enzyme (lane 2)*

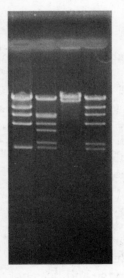

Molecular biology is **the** field these days. It is where the money is, both commercially and academically. All that we know about DNA and molecular biology came from the discovery of restriction enzymes. It seems so simple: Just cut up some DNA! It is such a powerful tool because of the diverse nature of the many restriction enzymes. Besides classical basic research, restriction enzymes are used in many fascinating ways. Forensic medicine uses restriction digestion of samples to identify people. Many criminals have been caught and identified by the specific restriction digest pattern of their DNA. Some states currently offer prison inmates free DNA tests to prove their claims of innocence. In one bizarre case, a prisoner was released after his DNA test showed that he was not guilty of the rape for which he had been imprisoned; however, he was rearrested a week later when his sample was shown to match that from three other unsolved cases.

Experiment 11

Analysis of DNA Restriction Fragments

In this experiment, you will use restriction enzymes to digest phage λ DNA and then run an agarose gel to separate the fragments. Using information from a known restriction enzyme, you will determine where on the phage λ DNA a second, unknown enzyme cuts.

Prelab Questions

1. What will you do immediately before having the restriction enzymes added to your reactions?

2. What are the safety techniques you will use with respect to use of ethidium bromide?

3. What is the purpose of using commercial *Hin*dIII fragments of λ DNA (lane 4) as well as the ones you generate yourself (lane 1)?

Objectives

Upon successful completion of this lab, you will be able to

- Conduct restriction digestion reactions.
- Prepare agarose gels and load DNA samples.
- Stain gels with ethidium bromide (EtBr) to visualize DNA.
- Analyze DNA fragments and determine the molecular weight of the fragments.
- Draw a partial restriction enzyme map.

In this experiment you will digest λ DNA with *Hin*dIII, *Mys*I (a mystery endonuclease), and a mixture of both. You will use the results obtained to make a restriction map.

Experimental Procedures

Materials

Phage λ DNA and commercial λ DNA/*Hind*III fragments

*Hind*III

*Mys*I

Reaction buffer for endonucleases

TAE buffer (Tris-Acetate-EDTA)

Bromophenol blue/zylene cyanol tracking dye

37°C and 65°C water baths or hot blocks

Electrophoresis equipment

DNA stain (EtBr will be added at the end)

Pipetmen and tips

Methods

Digesting λ DNA

1. You will receive microfuge tubes with λ DNA, reaction buffer, and commercial *Hind*III λ fragments.

2. Keep all your samples in an ice bucket when not incubating at a prescribed temperature.

3. Set up three reactions as shown but do not add the *Hind*III and *Mys*I.

Reaction	1	2	3
10× reaction buffer (μL)	1	1	1
dd H$_2$O (μL)	6	6	5
λ DNA, 0.5 μg/μL (μL)	2	2	2
*Hind*III (μL)	1	—	1
*Mys*I (μL)	—	1	1

4. Spin the tubes in a microfuge for a few seconds. Take them to the instructor or teaching assistant to have the enzymes added. Mix and spin again.

TIP 11.1   **Caution!** Anytime that you are anywhere near ethidium bromide, you must be properly dressed. This includes gloves, long sleeves, eye protection, and closed-toed shoes.

5. Incubate the tubes at 37°C for 20 min.

6. When the incubation is over, add 2 μL of bromophenol blue/xylene cyanol tracking dye to the three samples and to your tube of commercial λ fragments.

7. Place the four tubes in the 65°C water bath for 5 min and then return them to the ice bucket quickly.

Agarose Gel Electrophoresis

1. While the digestion is proceeding, make 40 mL of 1% agarose in TAE buffer for the submarine gels (see Experiment 9).

2. Level a casting tray, pour the gel, insert the comb, and let the gel harden for 15 min.

3. Carefully remove the rubber stopper and comb.

4. Pour the TAE reservoir buffer so that it is just over the gel. Make sure the positive pole is away from the wells.

5. Add your samples to the wells.

6. Set the voltage to 120.

7. When done electrophoresing, place the gel into a *clean* plastic container, taped and labeled. Take your gel into the predetermined room to have the EtBr added.

8. View the gel and have a picture taken.

9. Measure the migration of the bands on the picture.

TIP 11.2 **Danger!** Never stare directly at the UV lightbox or leave exposed flesh over it. It is much stronger than a tanning booth and can burn you quickly.

Name _____   Section _____

Lab partner(s) _____   Date _____

Analysis of Results

Experiment 11: **Analysis of DNA Restriction Fragments**

Data

1. Turn in your photograph or photocopy of the gel, labeling the lanes. If neither is available, draw a picture of the gel on a separate piece of paper.

2. Measure the distances of the bands, starting with the bands the highest up (closest to the wells). Record your findings in the following table.

Lane 1	Lane 2	Lane 3	Lane 4
_____	_____	_____	_____
_____	_____	_____	_____
_____	_____	_____	_____
_____	_____	_____	_____
_____	_____	_____	_____
_____	_____	_____	_____
_____	_____	_____	_____
_____	_____	_____	_____
_____	_____	_____	_____
_____	_____	_____	_____

Calculations

1. On a piece of graph paper, graph the log base pairs versus the migration distance for the *Hind*III fragments. Draw a line or smooth curve

TIP 11.3 When plotting the size of DNA fragments versus migration, don't force a straight line through the data. Very often your data fit better to a curve.

that really fits the data. If the points do not make a straight line, do not force them. Determine the base pairs for the unique *Mys*I fragments and double-digest fragments.

2. What fragments from one digest have a restriction site for the other endonuclease within them? (For example, the 6.5-kb *Hind*III fragment has a *Mys*I site within it.)

3. Using the known location of the *Hind*III sites and their fragment sizes, the calculated sizes of the *Mys*I fragments, and the information from the double digest, make a restriction map showing as many of the *Mys*I sites as you can. It is easiest if you include the *Hind*III sites on the top and the *Mys*I sites on the bottom of a line. If your results are not good, get a picture of a more appropriate gel that you can use for this question.

Questions

1. The restriction enzyme *Bst*XI recognizes the sequence CCANTGG (N stands for any nucleotide) whereas *Hha*I recognizes GCGC. Which of these two enzymes is more likely to cut a given piece of DNA most often? Explain.

2. A DNA fragment of 3300 base pairs was digested with *Eco*R1 alone, with *Hind*III alone, and with *Eco*RI and *Hind*III together. The fragments obtained and their sizes are shown in the figure. Draw a restriction map based on this information.

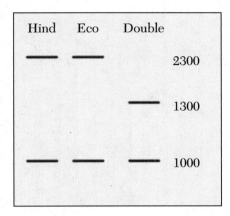

3. The restriction enzyme, *Xho*I, cuts phage λ DNA at position 33,498. Draw a picture of the gel you will see if it is used as the mystery enzyme in the experiment.

Additional Problem Set

1. What is the importance of methylation in the activity of restriction endonucleases?

2. Why do restriction endonucleases *not* hydrolyze DNA from the organism that produces them?

3. What is a palindrome? What English words or sentences are palindromes?

4. What is the importance of *sticky ends* in molecular biology?

5. The restriction enzyme, *Sst*I, cuts phage λ DNA at positions 24,472 and 25,877. Draw a picture of the gel you will see if it is used as the mystery enzyme in Experiment 11.

6. The restriction enzyme, *Xba*I, cuts phage λ DNA at position 24,508. Draw a picture of the gel you will see if it is used as the mystery enzyme in Experiment 11.

7. Which of the two restriction enzymes, *Xba*I, *or Sst*I, is best to use to run a 1.5% agarose gel instead of a 1% gel? (See the information in Problems 5 and 6.)

Webconnections

For a list of websites related to the material covered in this chapter, go to **Webconnections** at the *Experiments in Biochemistry* site on the Brooks/Cole Publishing website. You can access this page at http://www.brookscole.com and follow the links from the chemistry page.

References and Further Reading

Anderson, J. N. *A Laboratory Course in Molecular Biology.* Dayton, IN: Modern Biology, 1986.

Ausubel, F., R. Brent, R. Kingston, D. Moore, J. Seidman, J. Smith, and K. Struhl. *Short Protocols in Molecular Biology.* New York: Wiley, 1992.

Berg, P., and M. Singer. *Dealing with Genes: The Language of Heredity.* Mill Valley: University Science Books, 1992.

Bodmer, W., and R. McKie. *The Book of Man: The Human Genome Project and the Quest to Discover Our Genetic Heritage.* New York: Simon and Schuster, 1995.

Boyer, R.F. *Modern Experimental Biochemistry.* Menlo Park, CA: Addison-Wesley, 1993.

Campbell, M., and S. Farrell. *Biochemistry.* Belmont, CA: Brooks/Cole, 2005.

Dryer, R. L., and G. F. Lata. *Experimental Biochemistry.* New York: Oxford University Press, 1989.

Haseltine, W. A. "Discovering Genes for New Medicines." *Scientific American* 276, no. 3 (1997).

Lehninger, A. L., D. L. Nelson, and M. M. Cox. *Principles of Biochemistry,* 2nd ed. New York: Worth, 1993.

Pennisi, E. "Chemical Shackles for Genes?" *Science* 273 (1996).

Roberts, L. "New Scissors for Cutting Chromosomes." *Science* 249 (1990).

Robyt, J. F., and B. J. White. *Biochemical Techniques.* Long Grove, IL: Waveland Press, 1990.

Stryer, L. *Biochemistry*, 3rd ed. New York: Freeman, 1988.

Chapter 12

Cloning and Expression of Foreign Proteins

TOPICS

Introduction

*In this chapter, we discuss the methods used to recombine DNA from different sources, a procedure known as **cloning**. This is usually done with the ultimate goal of producing large quantities of a protein of interest, such as an important hormone or immunity factor. With cloning, the rapid proliferation of a bacterial strain can be used to produce much larger quantities of a mammalian protein than would be possible by traditional purification methods. Often a fusion protein is expressed, where the protein of interest is attached to an amino acid sequence that makes it easy to purify.*

12.1 Recombinant DNA

The entire field of molecular biology grew from the ability to manipulate genetic information using recombinant DNA technology. Recombinant DNA is DNA from two different sources that have been recombined so that a gene of interest can be studied. Recombinant DNA allows individual genes to be cloned so that researchers can study both the structure and function separately from the organism in which the genes were originally isolated. This chapter explains the process of cloning and expressing foreign proteins in bacteria. Figure 12.1 shows the basics of creating recombinant DNA.

Insulin was the first commercially available, genetically engineered product that was a direct result of expressing a foreign protein in bacteria. Since then, many products have been created using this technology, including some cancer-fighting drugs like interferon. The term *cloning* has many different meanings depending on the context in which it is used. Molecular biologists use the term to describe the process of creating recombinant DNA molecules. In microbiology, bacteria replicate by binary

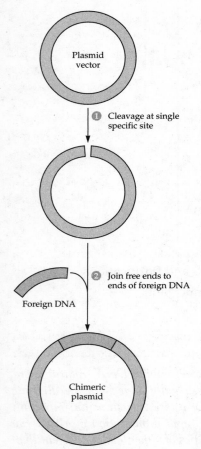

FIGURE 12.1 *Creating recombinant DNA*

fission giving rise to a colony of cells that are identical clones. A more strict definition of a clone is a group of genetically identical organisms. Perhaps the most common example of a clone is an identical twin.

Here are some terminologies that you should become familiar with as you read through this chapter. **Foreign DNA,** or **insert,** is the segment of DNA, usually a single gene, that is of interest to the researcher. This segment of DNA is usually cloned and contains a gene or at least part of a gene. Most of the time, the segment of DNA encodes a protein and is referred to as the **open reading frame (ORF).** To clone the segment of DNA and express it in bacteria, you must first put it into a carrier molecule, or vector (see Section 12.2). Recombinant DNA technology would not be possible without the discovery of **restriction endonucleases,** which we studied in Chapter 11. Another important enzyme is **DNA ligase,** which catalyzes the resealing of the DNA backbone once it has been cut by forming phosphodiester bonds between the nucleotides.

Once the foreign DNA, or insert, has been ligated into a vector, it is transferred to a host bacterial cell line via a process referred to as **transformation.** The host cell then replicates, producing identical offspring (clones), as well as replicating the foreign DNA. Thus, as long as the foreign gene can be replicated, we can clone the gene for as long as the host cell bacteria continue to clone themselves. This leads to a huge amplification of the cloned DNA because most bacteria divide every 20 minutes under optimal conditions.

Bacteria are very prolific, so finding the bacterial cells that actually were transformed is necessary; this process is called **selection.** The vector will have some type of marker that enables you to spot the bacterial cells that took up the vector with insert. Common selectable markers are genes for resistance to antibiotics.

Once there are large quantities of the cloned gene, the gene can be expressed. This means the gene is transcribed to mRNA, and then the mRNA is translated into protein. This is the process of **expression** and often requires a specific vector and bacterial cell line that have the ability to express foreign proteins. Some vectors have sequences upstream or downstream of the cloning site that allow the expression of the desired protein such that it is attached to a special amino acid sequence. The altered protein is called a **fusion protein.**

12.2 Vectors

A **vector** is a piece of DNA that can be used to carry foreign DNA into host cells. A vector, by definition, replicates autonomously (separately from chromosomal DNA) and can therefore be used as a cloning vehicle for identifying newly created recombinant DNA molecules. Once the foreign DNA is recombined with a vector, it is transferred to a host cell. The vector with the foreign DNA is replicated, producing clones that can be isolated and analyzed. The most common vectors are plasmids, bacteriophages, and cosmids. All can be used for cloning, but we consider only plasmids in this chapter.

Plasmids are small autonomously replicating pieces of circular DNA that were originally identified in bacteria because they contain genes that allow for resistance to certain antibiotics. It was then found that bacteria can pass around these antibiotic-resistance plasmids from cell to cell. The cell receiving the plasmid then becomes resistant to the antibiotic and can grow in high concentrations of the antibiotic. The plasmid's ability to confer antibiotic resistance is necessary for use in cloning because researchers are interested only in cells that contain the plasmid. However, one unfortunate consequence of plasmid transfer between bacteria is that more and more bacteria are becoming resistant to certain antibiotics. Figure 12.2 shows a typical plasmid, pGEM.

Plasmids have been modified to allow their use in cloning. Today's plasmids have all been designed with the researcher's needs in mind and do not resemble the original bacterial plasmids from which they were derived. To be used as a vector for cloning, a plasmid must have several features. For the plasmid to be replicated independently of the host genome, it must have its own **origin of replication.** This is usually indicated on a plasmid map as **oriC.** A plasmid must have a **selectable marker,** such as a gene that confers resistance to an antibiotic. The presence of a selectable marker allows for growth of *Escherichia coli* in the presence of an antibiotic, thus allowing selection of cells that contain the plasmid. The plasmid map shown in Figure 12.2 has one selectable marker, the gene for ampicillin-resistance, Amp^r. The plasmid must also have many unique **restriction sites,** sites for cleavage by restriction enzymes (Chapter 11) for use when inserting the foreign DNA into the plasmid. These unique sites are often concentrated on a region of high restriction–site density called a **multiple-cloning site** (MCS). With the pGEM vector shown in Figure 12.2,

FIGURE 12.2 *pGEM plasmid vector*

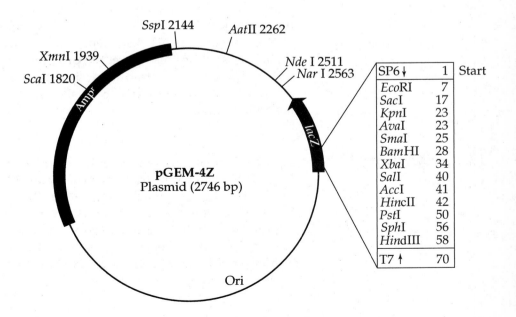

the MCS is within the *lacZ* gene and has 13 different restriction enzyme sites in 58 base pairs.

The purpose of having the *lacZ* gene in the MCS is that it allows for easy screening (**blue/white screening**) of the transformed cells that have the foreign DNA. When the MCS is within the *lacZ* gene, insertion of foreign DNA will disrupt the gene, rendering it nonfunctional. If for some reason the foreign DNA is not inserted into the MCS, the host cell will then produce the β-galactosidase protein from the *lacZ* gene. To screen for insertion of foreign DNA when cloning, an *E. coli* strain that contains a nonfunctional β-galactosidase protein is used. The recombinant plasmid is transformed into this strain, and the resulting transformants are grown on plates containing a dye called X-gal, which turns blue when the β-galactosidase protein is present. Thus, if an *E. coli* cell contains a vector with foreign DNA in the MCS, it will remain white. If the vector does not contain the insert, then the cells will turn blue because they are producing the β-galactosidase protein. This allows a quick distinction between cells that took up a plasmid with the insert and those that took up a plasmid without the insert.

12.3 Foreign DNA

Most of the time, the foreign DNA inserted in a cloning experiment is a gene encoding a single polypeptide. This foreign DNA can be obtained using any number of standard techniques (for example, polymerase chain reaction). Once the DNA is cloned into a vector, it can then be used for many different applications—for example, expressing and purifying the protein in *E. coli*. If the gene encodes an unknown protein, then the researcher can sequence the DNA and deduce (using the genetic code) the amino acid sequence of the protein. This type of analysis can be very useful when trying to understand the function of the protein. Computer programs can also analyze the sequence. These programs, for example, can search for restriction sites, open reading frames, and percent homology to other genes. Figure 12.3 shows a printout from a DNA analysis program (DNA Strider 1.2). The sequence of barracuda lactate dehydrogenase (LDH) is listed. In this figure, the program was instructed to find the first start codon and begin translating it until it encounters a stop codon. The translated amino acid sequence is shown using the single letter codes for the amino acids.

12.4 Restriction Enzymes

Figure 12.4 gives an example of how a DNA analysis program can analyze DNA sequences for restriction sites. Figure 12.5 shows a complete restriction map indicating which enzymes cut the barracuda LDH open reading frame, what the sites are, and where they cut. Another useful piece of information is knowing the enzymes that don't cut the DNA. This information is useful when trying to clone the segment of DNA. The DNA Strider program also gives the list of restriction sites *not* found in the DNA, as shown in Figure 12.6.

```
   1  GGAAAAGCTGCACTCAGGAGGAGACACATTCCAGCTCGCGCCTTCGCCTTTATCTTAAAAACACCTCCCTCCCAAAAAAAAGAAAAAAAAACCTCAAG    98

  99  ATG TCC ACC AAG GAG AAG CTC ATC GAC CAC GTG ATG AAG GAG GAG CCT ATT GGC AGC AGG AAC AAG GTG ACG GTG  173
   1   M   S   T   K   E   K   L   I   D   H   V   M   K   E   E   P   I   G   S   R   N   K   V   T   V    25

 174  GTG GGC GTT GGC ATG GTG GGC ATG GCC TCC GCC GTC AGC ATC CTG CTC AAG GAC CTG TGT GAC GAG CTG GCC CTG  248
  26   V   G   V   G   M   V   G   M   A   S   A   V   S   I   L   L   K   D   L   C   D   E   L   A   L    50

 249  GTT GAC GTG ATG GAG GAC AAG CTG AAG GGC GAG GTC ATG GAC CTG CAG CAC GGA GGC CTC TTC CTC AAG ACG CAC  323
  51   V   D   V   M   E   D   K   L   K   G   E   V   M   D   L   Q   H   G   G   L   F   L   K   T   H    75

 324  AAG ATT GTT GGC GAC AAA GAC TAC AGT GTC ACA GCC AAC TCC AGG GTG GTG GTG GTG ACC GCC GGC GCC CGC CAG  398
  76   K   I   V   G   D   K   D   Y   S   V   T   A   N   S   R   V   V   V   V   T   A   G   A   R   Q   100

 399  CAG GAG GGC GAG AGC CGT CTC AAC CTG GTG CAG CGC AAC GTC AAC ATC TTC AAG TTC ATC ATC CCC AAC ATC GTC  473
 101   Q   E   G   E   S   R   L   N   L   V   Q   R   N   V   N   I   F   K   F   I   I   P   N   I   V   125

 474  AAG TAC AGC CCC AAC TGC ATC CTG ATG GTG GTC TCC AAC CCA GTG GAC ATC CTG ACC TAC GTG GCC TGG AAG CTG  548
 126   K   Y   S   P   N   C   I   L   M   V   V   S   N   P   V   D   I   L   T   Y   V   A   W   K   L   150

 549  AGC GGG TTC CCC CGC CAC CGC GTC ATC GGC TCT GGC ACC AAC CTG GAC TCT GCC CGT TTC CGC CAC ATC ATG GGA  623
 151   S   G   F   P   R   H   R   V   I   G   S   G   T   N   L   D   S   A   R   F   R   H   I   M   G   175

 624  GAG AAG CTC CAC CTC CAC CCT TCC AGC TGC CAC GGC TGG ATC GTC GGA GAG CAC GGA GAC TCC AGT GTG CCT GTG  698
 176   E   K   L   H   L   H   P   S   S   C   H   G   W   I   V   G   E   H   G   D   S   S   V   P   V   200

 699  TGG AGT GGA GTG AAC GTT GCT GGA GTT TCT CTG CAG ACC CTT AAC CCA AAG ATG GGG GCT GAG GGT GAC ACG GAG  773
 201   W   S   G   V   N   V   A   G   V   S   L   Q   T   L   N   P   K   M   G   A   E   G   D   T   E   225

 774  AAC TGG AAG GCG GTT CAT AAG ATG GTG GTT GAT GGA GCC TAC GAG GTG ATC AAG CTG AAG GGC TAC ACT TCC TGG  848
 226   N   W   K   A   V   H   K   M   V   V   D   G   A   Y   E   V   I   K   L   K   G   Y   T   S   W   250

 849  GCC ATC GGC ATG TCC GTG GCT GAC CTG GTG GAG AGC ATC GTG AAG AAC CTG CAC AAA GTG CAC CCA GTG TCC ACA  923
 251   A   I   G   M   S   V   A   D   L   V   E   S   I   V   K   N   L   H   K   V   H   P   V   S   T   275

 924  CTG GTC AAG GGC ATG CAC GGA GTA AAG GAC GAG GTC TTC CTG AGT GTC CCT TGC GTC CTG GGC AAC AGC GGC CTG  998
 276   L   V   K   G   M   H   G   V   K   D   E   V   F   L   S   V   P   C   V   L   G   N   S   G   L   300

 999  ACG GAC GTC ATT CAC ATG ACG CTG AAG CCC GAA GAG GAG AAG CAG CTG GTG AAG AGC GCC GAG ACC CTG TGG GGC 1073
 301   T   D   V   I   H   M   T   L   K   P   E   E   E   K   Q   L   V   K   S   A   E   T   L   W   G   325

1074  GTA CAG AAG GAG CTC ACC CTG TGA GTGTCGCTCCTCTGATTTCTCCAGTCCGCCCTGAAAACACACCAAACACTGTGTGGTTATCCCCTCCC 1165
 326   V   Q   K   E   L   T   L   *                                                                      333
```

FIGURE 12.3 *Barracuda LDH-A gene sequence printout by the DNA Strider computer program*

Once you have all the information on restriction enzymes and the sequences of the foreign DNA and vectors, you can plan the cloning strategy. To clone foreign DNA into a suitable vector, you need to find unique restriction sites that are outside the actual gene itself (ORF). If there are no convenient sites, then you can mutate or change the DNA sequence to introduce one. Remember, this won't affect the protein because you won't change the ORF. For example, if you want to clone the DNA shown in Figures 12.3 and 12.4, do not pick *Alu*I to cut out the gene. Although *Alu*I has a site at position 33, which is upstream of the translation start site (the ATG at position 99), an *Alu*I site is also at position 115. Thus, if you choose that restriction enzyme, you will cut the coding region of the gene into pieces.

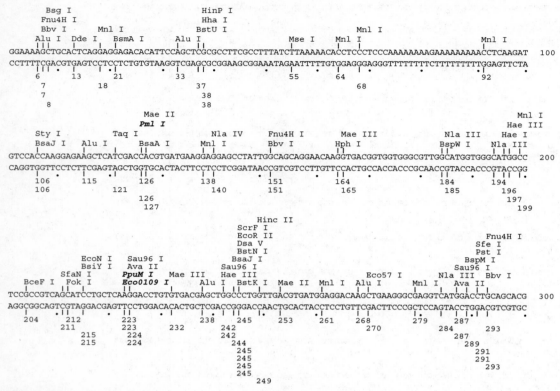

FIGURE 12.4 *Partial restriction map of barracuda LDH-A from the DNA Strider program*

Instead, choose another restriction enzyme that does not cut into the coding region. You also need to find the same restriction site in the vector in which you are planning to clone the foreign DNA. Cloning this DNA into the pGEM vector shown in Figure 12.2 is difficult because no common restriction sites are available between the vector and the upstream noncoding region. (Chapter 13 explains how this problem might be avoided.)

Another consideration is the number of restriction enzymes to use. If you have a DNA insert cut out with *Eco*RI at both ends, ligating this into a vector that was opened up with the same enzyme is easy (see Figure 12.1). An important technical consideration is how to keep the vector from simply resealing because *Eco*RI leaves sticky ends that are self-complimentary. Do this by removing the 5′ phosphate group on the end of the DNA with the enzyme phosphatase. As the name implies, phosphatase is specially suited to remove phosphate from DNA. On the other hand, if the foreign DNA is cut with two different enzymes and the vector is cut with the same enzymes, you can ligate the foreign DNA into the vector without worrying about the vector reclosing because the sticky ends would not match. Figure 12.7 shows directional cloning using two restriction enzymes.

FIGURE 12.5 *List of restriction enzymes and sites for barracuda LDH-A from the DNA Strider program*

Enzyme		Site	Use	Site position	(Fragment length)	Fragment order
Alw	I	ggatc 4/5	1	1(660) 2	661(1649)	1
AlwN	I	cagnnn/ctg	1	1(1378) 1	1379(931)	2
Ase	I	at/taat	1	1(2175) 1	2176(134)	2
Avr	II	c/ctagg	1	1(1991) 1	1992(318)	2
Ban	II	grgcy/c	1	1(1082) 2	1083(1227)	1
Bbe	I	ggcgc/c	1	1(386) 2	387(1923)	1
Bbs	I	gaagac 2/6	1	1(956) 2	957(1353)	1
Bcl	I	t/gatca	1	1(819) 2	820(1490)	1
Bgl	II	a/gatct	1	1(2161) 1	2162(148)	2
BstE	II	g/gtnacc	1	1(376) 2	377(1933)	1
BstY	I	r/gatcy	1	1(2161) 1	2162(148)	2
Bsu36	I	cc/tnagg	1	1(1204) 1	1205(1105)	2
Drd	I	gacnnnn/nngtc	1	1(341) 2	342(1968)	1
Eae	I	y/ggccr	1	1(1728) 1	1729(581)	2
Ecl136	I	gag/ctc	1	1(1082) 2	1083(1227)	1
Eco0109	I	rg/gnccy	1	1(222) 2	223(2087)	1
Ehe	I	ggc/gcc	1	1(386) 2	387(1923)	1
Esp	I	gc/tnagc	1	1(544) 2	545(1765)	1
Fsp	I	tgc/gca	1	1(2046) 1	2047(263)	2
Gdi	II	yggccg -5/-1	1	1(1728) 1	1729(581)	2
Kas	I	g/gcgcc	1	1(386) 2	387(1923)	1
Nae	I	gcc/ggc	1	1(383) 2	384(1926)	1
Nar	I	gg/cgcc	1	1(386) 2	387(1923)	1
Nco	I	c/catgg	1	1(1350) 1	1351(959)	2
PflM	I	ccannnn/ntgg	1	1(610) 2	611(1699)	1
Pml	I	cac/gtg	1	1(125) 2	126(2184)	1
PpuM	I	rg/gwccy	1	1(222) 2	223(2087)	1
PshA	I	gacnn/nngtc	1	1(997) 2	998(1312)	1
Sac	I	gagct/c	1	1(1082) 2	1083(1227)	1
SgrA	I	cr/ccggyg	1	1(382) 2	383(1927)	1
Spe	I	a/ctagt	1	1(2299) 1	2300(10)	2
Sph	I	gcatg/c	1	1(933) 2	934(1376)	1
Ssp	I	aat/att	1	1(1359) 1	1360(950)	2
Stu	I	agg/cct	1	1(301) 2	302(2008)	1
Tth111	I	gacn/nngtc	1	1(950) 2	951(1359)	1

No Sites found for the following Restriction Endonucleases

Acc	I	gt/mkac	BspE	I	t/ccgga	Mlu	I	a/cgcgt	Sap	I	gcttcttc 1/4
Afl	II	c/ttaag	BspH	I	t/catga	Msc	I	tgg/cca	Sca	I	agt/act
Afl	III	a/crygt	BssH	II	g/cgcgc	Nci	I	cc/sgg	Sfi	I	ggccnnnn/nggcc
Age	I	a/ccggt	BstB	I	tt/cgaa	Nde	I	ca/tatg	Sma	I	ccc/ggg
Apa	I	gggcc/c	Cla	I	at/cgat	Nhe	I	g/ctagc	Sna	I	gta/tac
Asp718		g/gtacc	Eag	I	c/ggccg	Not	I	gc/ggccgc	SnaB	I	tac/gta
Ava	I	c/ycgrg	Eco47	III	agc/gct	Nru	I	tcg/cga	Spl	I	c/gtacg
BamH	I	g/gatcc	EcoR	I	g/aattc	Nsi	I	atgca/t	Sse8337	I	cctgca/gg
Bcn	I	ccs/gg	EcoR	V	gat/atc	Pac	I	ttaat/taa	Swa	I	attt/aaat
Bgl	I	gccnnnn/nggc	Fse	I	ggccgg/cc	PaeR7	I	c/tcgag	Xba	I	t/ctaga
BsaB	I	gatnn/nnatc	Hind	III	a/agctt	Pvu	I	cgat/cg	Xca	I	gta/tac
BsiE	I	cgry/cg	Hpa	I	gtt/aac	Rsr	II	cg/gwccg	Xho	I	c/tcgag
Bsm	I	gaatgc 1/-1	Kpn	I	ggtac/c	Sac	II	ccgc/gg	Xma	I	c/ccggg
Bsp120	I	g/ggccc	Mcr	I	c/grycg	Sal	I	g/tcgac	Xmn	I	gaann/nnttc

FIGURE 12.6 *Restriction sites* not *found in barracuda LDH-A by the DNA Strider program*

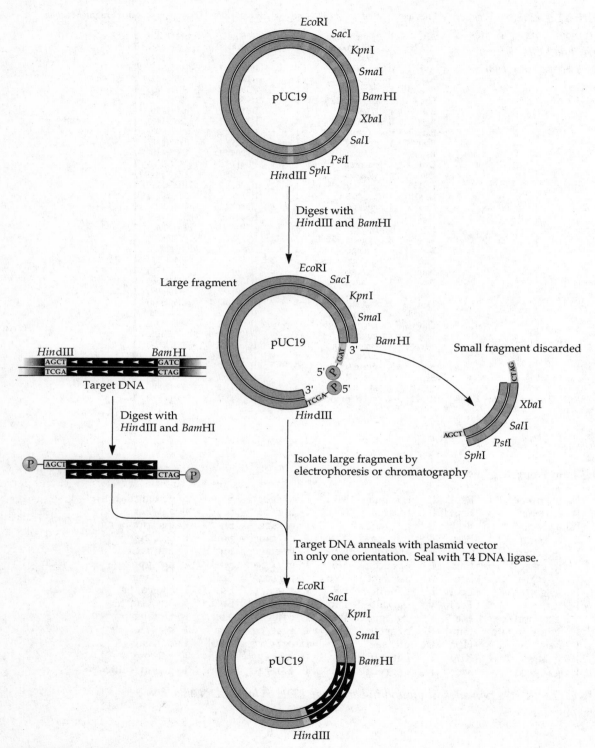

FIGURE 12.7 *The basics of directional cloning. Two restriction enzymes are used so that the inserted DNA can go into the plasmid in only one orientation.*

PRACTICE SESSION 12.1   If you have the following gene and want to clone it into the pGEM vector, what restriction enzymes will you choose? The restriction map of an example gene is also shown. The start codon (ATG) is at position 99; the stop codon (TGA) is at position 1095.

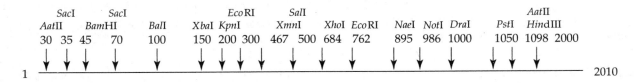

Restriction Enzyme Usage

Enzyme	Position	Enzyme	Position
*Aat*II	30,2000	*Not*I	986
*Bal*I	100	*Pst*I	1050
*Bam*HI	45	*Sac*I	70
*Dra*I	1000	*Sal*I	500
*Eco*RI	300,762	*Sca*I	35
*Hind*III	1098	*Xba*I	150
*Kpn*I	200	*Xho*I	684
*Nae*I	895	*Xmn*I	467

The strategy is to find two restriction sites that appear inside the MCS of the pGEM vector and that surround the coding region of the gene. This can be done by trial and error, looking at each available restriction enzyme. For example, suppose we pick *Aat*II. That enzyme cuts at position 30, which is before the start codon at position 99. *Aat*II also has a site at position 2262, but that is outside the MCS (Figure 12.2). That is a bad choice if you want to use the *lac*Z gene for screening. *Sca*I cuts at position 35 on the gene but also cuts inside the ampicillin-resistance gene on the plasmid (position 1820). *Sac*I cuts at position 70 on the gene, which is before the start codon. It also cuts inside the MCS of the vector. On the other end, we need a restriction enzyme that cuts after the stop codon. Only two restriction enzymes listed cut after the stop codon at 1095: *Hind*III and *Aat*II, the latter of which we have already eliminated. *Hind*III has a site inside the MCS. Thus, the best two enzymes to use are *Sac*I and *Hind*III.   ●

12.5 Cell Lines

A **cell line,** or **strain,** is a specific type of cell, either prokaryotic or eukaryotic, with a defined genotype. When using recombinant DNA technology to clone and express foreign proteins, two types of bacterial cell lines are used—those that are used for cloning foreign DNA and those that are used for expressing foreign proteins.

Each type of cell line has been modified to perform a certain function. In addition, most cell lines are designed to use special vectors that allow them to perform that function. For example, the bacterial strain BL21 (Novagen) is designed for the sole purpose of expressing foreign proteins. This strain contains a chromosomal copy of the gene for T7 RNA polymerase. This gene is under the control of the *lac*UV5 promoter, which means that the gene can be turned on or off by the addition of a certain inducer. To express proteins in this strain, put the foreign DNA into a specially designed vector that has the T7 promoter in front of the MCS. Once the DNA is inserted into the vector, the plasmid is transformed into the BL21 strain, and the protein is expressed by the addition of inducer. The addition of inducer turns on the T7 RNA polymerase gene and causes the production of T7 RNA polymerase, which recognizes and binds the T7 promoter and transcribes the foreign gene. The mRNA produced from this transcription is then translated into protein.

When doing cloning experiments, the type of cell line used for the recovery of the recombinant plasmid is critical. These types of cloning strains have been modified to ensure high yield and efficient recovery of recombinant plasmid. In addition, some have been modified to allow for screening of recombinant plasmid. For example, if you want to do blue/white screening as described in Section 12.2, you not only need a plasmid with the MCS in the *lac*Z gene but also need a host strain that is deficient in the α-subunit of β-galactosidase.

12.6 Transformation

Transformation is the process by which there is a heritable change in a cell or an organism brought about by exogenous DNA. In molecular biology, the exogenous DNA is usually a plasmid. Scientists have developed several methods to optimize the uptake of DNA by bacteria. This optimization process is referred to as "making the bacteria competent." The most common procedure for making bacteria competent involves incubation with $CaCl_2$ followed by mild heat. This makes the bacterial membrane porous enough to allow foreign DNA to enter the cytoplasm. Another way is called *electroporation*, where the bacterial cells are subjected to an electrical current, which also causes the cells to take up foreign DNA.

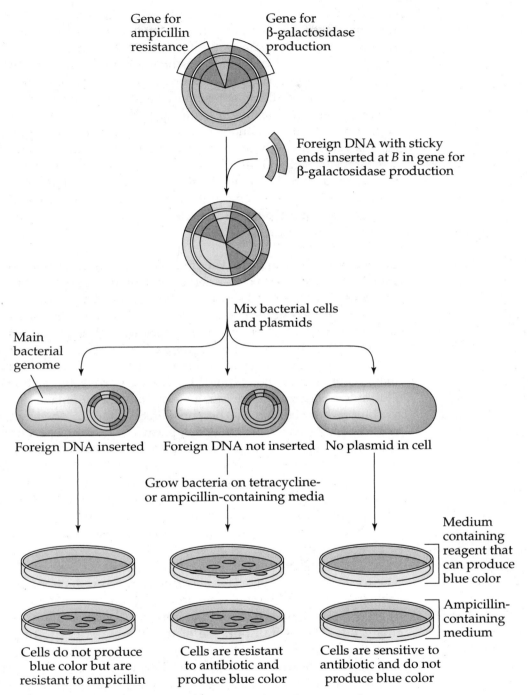

Gene for ampicillin resistance

Gene for β-galactosidase production

Foreign DNA with sticky ends inserted at *B* in gene for β-galactosidase production

Mix bacterial cells and plasmids

Main bacterial genome

Foreign DNA inserted   Foreign DNA not inserted   No plasmid in cell

Grow bacteria on tetracycline- or ampicillin-containing media

Medium containing reagent that can produce blue color

Ampicillin-containing medium

Cells do not produce blue color but are resistant to ampicillin

Cells are resistant to antibiotic and produce blue color

Cells are sensitive to antibiotic and do not produce blue color

FIGURE 12.8 *Cloning, transformation, and selection of recombinants, using ampicillin and blue/white screening*

Transformations, among other applications, are used when cloning foreign DNA into a vector (ligation reaction). Figure 12.8 shows a schematic of a transformation from a ligation reaction. This ligation mixture is simply incubated with competent cells and then plated out on selectable media. Because the efficiency of a ligation reaction is low, only a small percentage of the vectors will ligate to the foreign DNA, forming a circular piece of DNA. These new recombinant plasmids will be taken up by the competent cells. Because bacteria rarely take up linear DNA, most of the cells will not transform. These cells do not contain Amp^r gene and thus will not grow in the presence of ampicillin. In Figure 12.8, these are shown on the right; the cells shown on the left and center are transformed. Both took up a form of the plasmid, and both grow in the presence of ampicillin. The left-side cells were transformed with the plasmid carrying the insert. The center cells were transformed with the plasmid that simply resealed without the insert. (*Note:* This can happen only if the vector has compatible ends or the original digestion of the plasmid failed.)

12.7 Selection

Selection is the process of determining which bacterial cells were correctly transformed with the vector–insert combination; the selectable markers make this possible. Figure 12.8 shows one type of selection. If we assume that we have three types of cells in our mixture—the ones with no plasmid, the ones with recircularized plasmid, and the ones with plasmid and insert—we can plate this mixture on a medium of ampicillin. The cells that don't take up any plasmid will not grow on ampicillin because they do not carry the resistance that is conferred by the plasmid. These are shown as the plates on the right. Plasmids incorporating the *lacZ* gene are designed to make selection simple via blue/white screening, as described in Section 12.2. The foreign DNA disrupts the *lacZ* gene. A recircularized plasmid, however, has a functional *lacZ* gene. If the cells that are transformed lack the α-subunit of β-galactosidase, transformation by the recircularized plasmid restores activity. If the transformation mixture is plated onto a medium containing ampicillin and the dye X-gal, we can quickly identify the colonies that we want. They will be the white ones, as shown on the left of Figure 12.8. The blue colonies (shown in the middle) are the ones with the recircularized plasmid.

12.8 Expression

The focus of this chapter is the **expression,** or production, of foreign proteins in bacteria. So far, we have described how to clone foreign DNA into special expression vectors and the important considerations needed when doing this. The overall purpose of this cloning is to produce the protein that the DNA encodes. As described previously, special cell lines and expression vectors are used. Figure 12.9 shows a common expression vector called pET 5a. As is the case with any plasmid, the pET 5a vector has an origin of replication and a **selectable** marker (Amp^r). It also has an MCS, which is on the right side of the circular map. In front of this MCS is the T7 polymerase promoter, which allows for transcription (the production of mRNA) of any gene that is placed in the MCS. Thus, in the presence of T7 RNA polymerase, the gene in the MCS will be transcribed and translated into protein.

FIGURE 12.9 *pET 5a expression vector*

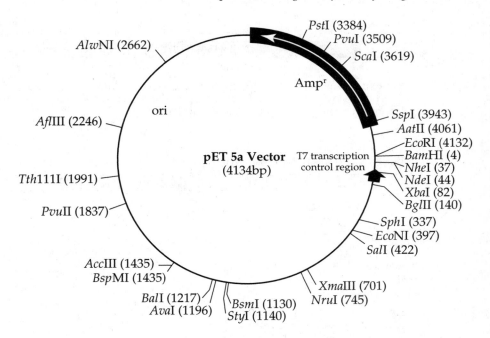

This description of expression is an oversimplification because many considerations are required for successful expression. First, some proteins are toxic when expressed in *E. coli*. You don't want these proteins produced all the time, so it is desirable to turn on the expression of the gene (foreign DNA) only when necessary. The expression is regulated by controlling the production of T7 RNA polymerase in the host cell, which is itself under the control of a lactose inducible promoter (*lac* operon). If you control the amount of T7 polymerase, you can control the transcription from the T7 promoter. The *lac* operon produces β-galactosidase when the cell needs to use the lactose as an energy source. A repressor protein is normally bound to the promoter preventing the host cell's RNA polymerase from transcribing the operon. Lactose is the natural inducer that binds the repressor, removing it from the promoter and allowing transcription. Figure 12.10 gives an overview of the *lac* operon. The molecule isopropyl-β-D-thiogalactopyranoside (IPTG) is a nonmetabolisable analog of lactose. When this molecule is added to culture medium, it binds and removes the *lac* repressor, and the system is induced.

Thus, when you induce the cells with IPTG, you induce the production of T7 RNA polymerase. This polymerase then finds the T7 promoter on the expression vector and transcribes the gene (foreign DNA) that you inserted. The host cell then translates the mRNA into protein. This system keeps the host cells from producing the foreign proteins until you induce them.

Each expression experiment needs to be optimized for different proteins being expressed because every protein is different. The concentration of IPTG needed for maximum induction must be determined. The IPTG can be added to the host cells at different time points in their

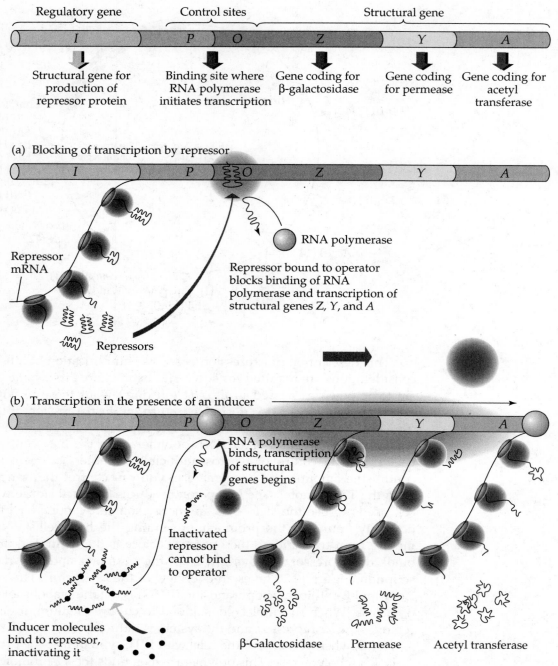

FIGURE 12.10 *The* lac *operon*

growth curve. The toxicity of the proteins produced must be checked. If the proteins kill the cells, you want to get maximum expression before the cells die. The time and temperature of induction is also a critical parameter.

A second consideration for expression is the orientation of the inserted DNA. If you use *Nde*I at the 5′ end and *Xba*I at the 3′ end to cut out a piece of foreign DNA, you can insert it into the pET vector, using those sites (see Figure 12.9). However, this puts the inserted DNA into the pET vector backward relative to the direction of transcription. If you use *Nde*I and *Eco*RI, then the insert goes into the pET vector in the proper orientation.

Remember that the ultimate goal is the isolation of large quantities of a protein that interests you. Figure 12.11 shows how cloning and expression have been used to produce large quantities of insulin, which was formerly very expensive and difficult to produce.

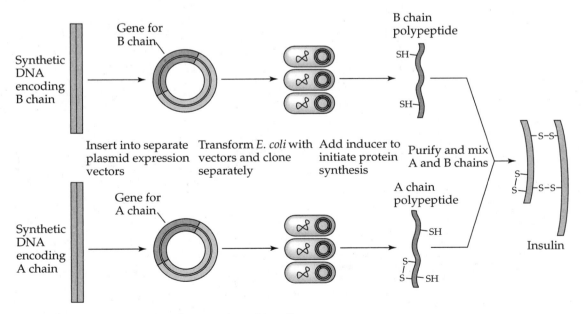

FIGURE 12.11 *Cloning and expression of insulin*

12.9 Fusion Proteins and Purifications

The expression of foreign proteins, a major breakthrough in molecular biology, was made possible by the use of expression vectors. Although large quantities of proteins were produced in a suitable cell line before this breakthrough, these proteins still needed to be separated and purified—a slow process—from the bacterial cells. The use of a newer type of expression vector has allowed a more rapid purification of the expressed proteins. These vectors produce a **fusion protein**.

Figure 12.12 shows the pET 15b plasmid. This vector is 5.7 kilobase (kb) in size and has the usual complement of features for an expression vector. The thick black arrow pointing counterclockwise shows the direction of expression in the area of the MCS. Note that there is an *Nde*I site and a *Bam*HI site. If a gene to be expressed can be isolated and has these two restriction sites, with the *Nde*I being on the 5' end of the coding sequence, then the gene can be inserted in the MCS and can be expressed.

Figure 12.13 shows an expansion of the MCS and expression region. Just upstream from the *Nde*I site is a sequence labeled His-Tag. The gene of interest is inserted downstream of this, using *Nde*I and *Bam*HI, for example. When expressed, the mRNA produced will include the sequence from the T7 promoter to the T7 terminator. When this mRNA is translated, the

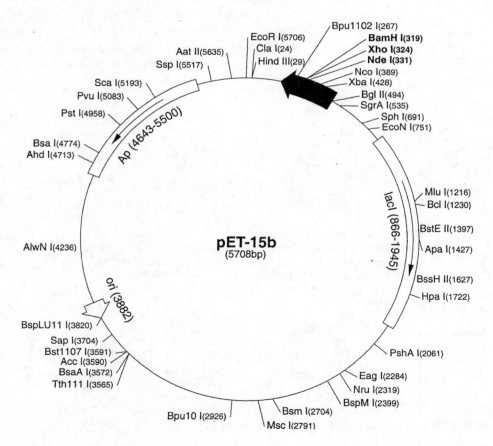

FIGURE 12.12 *The pET 15b vector from Novagen. The multiple cloning site is within a region that can be expressed.*

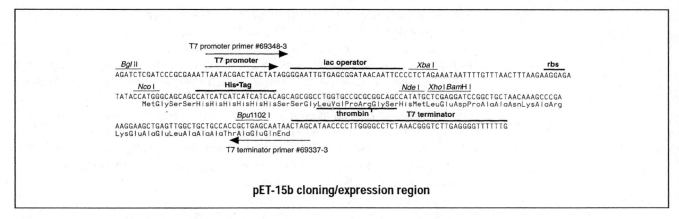

pET-15b cloning/expression region

FIGURE 12.13 *An expansion of the sequence in the multiple cloning site of pET 15b. There is a promoter for T7 RNA polymerase and a T7 terminator. There is also a sequence that puts a His-Tag onto the expressed protein.*

protein produced will have a series of six histidines at the N-terminal end of the protein before the protein encoded by the gene that was cloned.

The His-Tag is an important part of the expression because the fusion protein produced is easy to purify. The side chain of histidine is attracted to Ni^{2+}. If an affinity chromatography column is constructed with a nickel resin (see Chapter 6), then a protein with six histidines at the N-terminus will bind to the column. This procedure is shown in Figure 12.14. The cells

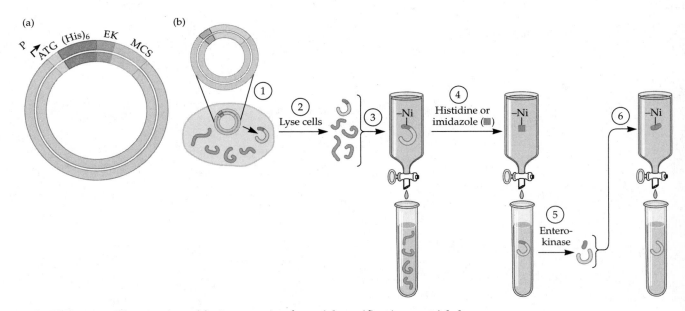

FIGURE 12.14 *The creation of fusion proteins for quick purification on nickel affinity columns*

that are expressing the desired protein are lysed, and the cell homogenate is run over the nickel column. The fusion protein sticks to the column, but the other proteins produced by the cells do not. Once the column is sufficiently washed, the fusion protein is eluted with imidazole, a compound similar to the side chain of histidine. Once the fusion proteins are collected, the His-Tag can be removed from the fusion protein with the enzyme enterokinase, which cleaves a sequence that is between the His-Tag and the N-terminus of the desired protein. Sometimes it is not necessary to remove the His-Tag. The fusion protein may be fully functional with the His-Tag still on. In other cases, the His-Tag may have to be removed to have the desired protein fully functional.

12.10 Why Is This Important?

Millions of people suffer from diabetes, a condition in which too little insulin is produced. Previously, insulin was only produced by purifying it from mammalian sources. This not only was expensive and time-consuming but also required the use of many laboratory animals. With the onset of cloning technology, insulin can now be produced using the replicative machinery of bacteria. It is cheaper and easier to obtain, improving the lives of those afflicted with the disease. As a pure science tool, cloning is used to produce and study DNA and protein sequences. Scientists are also attempting to use cloning techniques to repair damaged genes that cause a wide variety of health problems. This is known as gene therapy, a current and sometimes controversial topic.

Experiment 12

Cloning and Expression of Barracuda LDH-A

In this experiment, you will isolate barracuda LDH DNA and clone it into a pET 15b expression vector. An expression cell line is then transformed and the LDH expressed. This vector produces a His-Tag fusion protein with LDH, which can then be purified with a nickel affinity column.

This procedure starts with the barracuda LDH-A gene, which makes the M subunit of LDH (see Chapter 4). It was cloned into pCRScript by Dr. Linda Holland of the Scripps Oceanic Institute. Figure 12.15 shows a diagram of the plasmid and the LDH, noting the restriction sites for *Nde*I and *Bam*HI, which are used to cut out the LDH gene from the plasmid. The expression vector is pET 15b. In this case, the initial pET 15b plasmid already contains an insert, which is the rat kidney glutaminase gene (rKGA). This is cut out of pET 15b, using *Nde*I and *Bam*HI.

This long experiment is divided into several major parts. Your instructor may choose to omit certain parts or to have the teaching staff do certain parts for you. Some of the days in the lab are long, and some are very short.

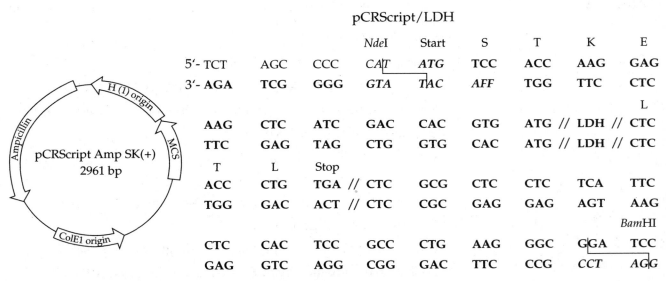

FIGURE 12.15 *pCRScript and LDH. This picture shows the sequence at the beginning and end where the LDH gene is inserted into the vector. The NdeI and BamHI restriction sequences are shown in italics.*

Prelab Questions

1. What is meant by the term *cloning*?

2. What is meant by the term *expression*?

3. What are the two vectors used in this experiment?

4. What are the two *E. coli* cell lines used?

5. Why are we using two different *E. coli* cell lines?

Objectives

The purpose of this experiment is to clone and express the gene for LDH from the barracuda, *Sphyraena lucasana*. In the process, the following ideas and techniques will be used:

    Preparation of competent *E. coli* cells

    Preparation of L-broth agar/ampicillin plates (LB/amp plates)

    Digestion of plasmid (vector) and foreign DNA with restriction enzymes

    Agarose gel electrophoresis of DNA

    Purification of DNA bands from agarose gels

    Ligation of foreign and vector DNA

    Transformation of competent cells and selection of transformants

    Induction of protein expression

    Purification of LDH fusion protein with affinity chromatography

    Verification of induction by SDS-PAGE and activity assays

This experiment counts on many things you cannot control, such as whether the DNA ligase is still active, and many things you can control, such as whether you can successfully pipet 0.5 µL of a solution. This would be a good time to check your pipetting technique with quantities of 1 µL and smaller.

Experimental Procedures: Part I

This section includes preparation of LB/amp plates, preparation of competent cells, digestion of plasmids with restriction enzymes, agarose gel separation of restriction fragments, and gel purification of plasmids and LDH DNA. Note that all references to water when part of a restriction digestion or other enzyme reaction refer to very pure water, such as double distilled or nanopure. For cloning cell lines, DH5 α is specified, but BL21 or JM109 are just as good.

Day 1

Today you pour LB-amp plates, which will eventually be used to screen for cloning and protein expression.

Preparing Amp Plates

1. Use a 1-L flask to prepare 200 mL of L-broth agar. L-broth agar contains the following: 0.5% bactoyeast extract, 1% NaCl, 1% tryptone, and 1.5% agar.

2. Cover the flask with aluminum foil, place a piece of autoclave tape on the flask, and autoclave for at least 20 min on a liquid cycle.

3. When the flask is done autoclaving, place it into a 52°C waterbath. Swirl occasionally while cooling. Swirl *gently* to keep air bubbles to a minimum.

4. When the flask has cooled to the temperature of the waterbath (15–30 min), add 200 µL of ampicillin (100 mg/mL) and swirl. If you can touch the flask for a few seconds without burning your hand, the media is cool enough.

5. Pour about 20 mL in each plate. Pour all the plates you can from your batch, even though you may not need them yourself. Extras may come in handy later.

6. Let the plates sit at room temperature for 24 h; then place them upside down in a plastic bag and store in a dark refrigerator until needed. Ampicillin is light sensitive.

Day 2

Today you perform the initial procedures for transformation, using the Fermentas TransformAid system. There are many ways to transform cells, but this system is fast and easy. Your instructor may choose a different method.

1. Inoculate $3 \times 2\,mL$ of TransformAid C-medium with DH5 α-cells. Incubate the cultures overnight at 37°C.

2. Mix one of the overnight cultures 50:50 with glycerol and freeze at –80°C. Save for future cultures.

Day 3

Transforming Cells with PCRScript/LDH and pET 15b/rKGA

1. Put 1.5 mL of C-medium into a microfuge tube and warm to 37°C.

2. Add 150 μL of overnight culture to the tube.

3. Place in a 37°C shaking incubator for 20 min.

4. Place the tube on ice. Spin the tube at 4°C for 1 min at $14,000 \times g$. Discard the supernatant. Save the tube containing the pellet for step 6.

5. Mix T-solution A with T-solution B in a 1:1 ratio to give the final T-solution for the next steps.

6. Add 300 μL of T-solution to the tube from step 4 and *gently* resuspend.

7. Keep on ice for 5 min.

8. Spin again at 4°C for 1 min. Discard the supernatant.

9. To the saved pellets from step 8, add 120 μL of T-solution to the tube. Keep on ice for 5 min.

10. Aliquot 1 μL (10–100 pg) of PCRScript/LDH to a 1.5-mL microfuge tube.

11. Aliquot 1 μL (10–100 pg) of pET 15b/rKGA to a 1.5-mL microfuge tube.

12. Put both tubes from steps 10 and 11 on ice for 5 min, along with another empty control tube.

13. Add 50 μL of T-cells to each DNA tube and the rest to the control tube.

14. Put the three tubes on ice for 5 min.

15. Plate the DNA and control cell mixes on LB/amp plates. For the control plate, plate the whole mix on the plate. For the plasmid plates, do two plates each, one with 10 μL and one with 40 μL.

16. Incubate the five plates overnight at 37°C.

Preparing Competent Cells

In this part, you prepare some *E. coli* cells that will be used for transformation later. The cell line is DH5α, BL21, or JM109, which are good for cloning. Eventually, you will need competent JM109 DE3 or BL21 DE3, which are expression cell lines. These can be prepared at the same time, prepared later, or purchased.

1. Acquire 100 mL of *E. coli* cells and prepare as follows: Glycerol stocks of cells are grown overnight in 50 mL of L-broth media. These are then used to inoculate 1 L of L-broth and grown until the optical density at 600 nm is between 0.5 and 0.6. This takes about 3 h at 37°C from the time the 50 mL are added to the liter. They are then cooled to 4°C.

2. Spin the cells at $2000 \times g$ for 15 min at 4°C. Use sterile 500-mL centrifuge bottles.

3. *Gently* resuspend the pellet in 1 mL of ice-cold, sterile 0.1 M $CaCl_2$, by drawing up and down into a pipet. *Gently* add 24 mL of sterile 0.1 M $CaCl_2$ and resuspend by drawing up and down into a 25-mL pipet while keeping the bottle cold in an ice bath. Spin again as in step 2.

4. *Gently* resuspend the pellet in 25 mL of ice-cold, sterile 0.1 M $CaCl_2$ as before, again keeping the bottle cold. Incubate on ice for 20 min. Spin again as in step 2.

5. *Gently* resuspend the pellet in 4.3 mL of 0.1 M $CaCl_2$. Add 0.7 mL of cold glycerol.

6. Aliquot 200 µL of cells into sterile cryo tubes and freeze quickly in a dry ice–EtOH bath. Store at −80°C.

Day 4

Remove the plates from the incubator in the afternoon and put in a refrigerator. Plates should be parafilmed around the edge and stored upside down.

Day 5

Preparing Overnight Cultures

1. Recover the plates from the refrigerator.

2. Choose three colonies from the pET 15b/rKGA plates, and for each one add a loopful to 10 mL of L-broth in sterile Falcon tubes. Add 10 µL of 100 mg/mL ampicillin.

3. Choose three colonies from the PCRScript/LDH plates, and for each one add a loopful to 10 mL of L-broth in sterile Falcon tubes. Add 10 µL of 100 mg/mL ampicillin.

4. Place in a 37°C shaking incubator overnight. Use culture tubes with loose lids and secure the lids in place with tape.

Day 6

Today you prepare the two plasmids from the overnight cultures and set up restriction enzyme digestions.

Preparing Plasmid Preps

Many companies make plasmid prep kits. The Qiagen kit and the Promega Gene Wizard are two of the most popular and are described below. Your instructor may prefer a different kit, which will change the procedures. Also, these kits change protocols often, so by the time you read this the protocol may well have changed again. There are also quick-and-cheap ways to purify plasmids that do not require kits.

Using Qiagen Plasmid Prep Kits

1. Spin down 1.7 mL of the pCRScript/LDH overnight culture. Discard the supernatant and add another 1.7 mL of the overnight culture and spin again, effectively doubling the amount of culture spun down. Save the leftover overnight for later culture.

2. In a separate tube, do the same thing for the pET 15b/rKGA overnight culture.

3. Resuspend the bacterial pellets in 250 μL of buffer P1, which should have had RNase A added already.

4. Add 250 μL of buffer P2 and mix *gently* by inversion four to six times or until the solution becomes viscous. Incubate at room temperature for 5 min. Do not let this step go for more than 5 min.

5. Add 350 μL of chilled buffer N3, mix immediately and *gently* by inversion four to six times, and incubate on ice for 5 min.

6. Mix the samples again and centrifuge at maximum speed in a microfuge for 10 min. Remove the supernatants promptly. The supernatants should be clear.

7. Apply the supernatants from step 6 to the spin columns.

8. Centrifuge for 1 min. Discard the flow-through.

9. Add 0.5 mL of buffer PB to the columns and centrifuge for 1 min. Discard the flow-through.

10. Add 0.75 mL of buffer PE to the spin columns and centrifuge for 1 min. Discard the flow-through.

11. Centrifuge the columns for 1 min to remove the last traces of wash buffer.

12. Place the spin columns in clean 1.5-mL microfuge tubes and add 50 μL of buffer EB to the center of the columns. Let stand for 1 min and then spin for 1 min.

13. The solutions that flowed through into the microfuge tube contain your purified plasmids. Save the plasmids at −20°C.

Using Wizard Plasmid Prep Kits

1. Split 3 mL of each of your six O/N cultures into two microfuge tubes each, such that each tube has 1.5 mL. Save the leftover overnight cultures for later.

2. Spin the tubes at 14,000 × g for 2 min to pellet the cells.

3. Resuspend the pellet from one tube in 250 μL of cell resuspension solution. After resuspending one tube, combine the resuspension with the other tube from the same culture and resuspend. Now you should have only six tubes again, one from each overnight culture that you made.

4. Add 250 μL of cell lysis solution. Invert six times *gently*. Add 10 μL of alkaline protease solution. Invert six times to mix and incubate at room temperature for 5 min.

5. Add 350 μL of neutralization solution. Invert six times to mix.

6. Centrifuge for 10 min at room temperature.

7. Insert spin columns into the 2-mL collection tubes (one for each sample).

8. Remove the supernatant from the spin in step 6 into the spin column.

9. Centrifuge for 1 min at room temperature. Discard the flow-through and reinsert the column into the collection tube.

10. Add 750 μL of wash solution. Centrifuge for 1 min. Discard the flow-through and reinsert the column into the collection tube.

11. Add 250 μL of wash solution. Centrifuge for 2 min. Discard the flow-through.

12. Transfer the spin columns to sterile microfuge tubes.

13. Add 100 μL of nuclease-free water to the spin column. Centrifuge for 1 min. The flow-through contains your plasmid for the digestion reactions. Discard the column. Save the plasmid at −20°C.

Using Quick-and-Cheap Plasmid Preps

1. Using a P-1000 tip, transfer 1.5 mL of overnight cultures to 1.5-mL microfuge tubes. Spin down at 14,000 × g for 3 min. Decant the supernatants, making sure that the cells have pelleted on the bottom of the tube.

2. Resuspend the cells in 100 μL TE buffer (10 mM TRIS, 1 mM EDTA, pH 8.0), using a P-200 tip. Make sure that the pellets are completely resuspended.

3. Add 200 μL of 0.2 M NaOH, 1% SDS. Vortex on high for 5 s and put on ice for exactly 5 min. The 5-min incubation is critical. If incubated too long, the DNA will denature.

4. Add 150 μL of cold 5 M KOAc. Vortex upside down. Put back on ice for 5–10 min.

5. Centrifuge 5 min at 14,000 × g. Transfer the supernatants to a clean microfuge tube. The pellets should be white and smeared on the side of the tube. Try not to transfer any of this to the clean tube. The longer the pellet sits before transfer, the looser the pellet becomes.

6. Add 1 mL of ice-cold EtOH to the transferred supernatants. Incubate in a freezer for 10 min. Invert several times. Centrifuge for 15 min at 14,000 × g. Sometimes you cannot see a pellet after this spin, so put the tubes in the centrifuge facing the same direction so you will know where the pellets are.

7. At the end of the spin, you should have a small, white pellet. This is your DNA. Decant the EtOH, making sure not to lose the pellet. To get rid of the excess EtOH, place the open tube in a 37°C hot block for 10 min.

8. Resuspend in 15 μL of water. Add the water to the side of the tube and close. Flick down the water and let it sit for 5 min.

Note: This prep will have much RNA and protein in it. It can be used for transformations and digestions. For digestions, RNase A should be added to the reaction.

Plasmid Digests

In this part, you cut the purified plasmids with *Nde*I and *Bam*HI. Be sure to save all plasmid preps and anything else that is leftover because sometimes they will be used for other things, including recovering from problems you may encounter.

You should have six samples to use. You will double-digest three samples each of your two different plasmids. Save 10 μL of each of your two plasmids to use as an undigested control later on.

1. Set up the digest protocol shown in the following table.

	Tube					
Reagent	*1*	*2*	*3*	*4*	*5*	*6*
pCRScript/LDH	10 μL	10 μL	10 μL			
pET 15/rKGA				10 μL	10 μL	10 μL
10× *Bam*HI buffer	2 μL	2 μL	2 μL	2 μL	2 μL	2 μL
BSA (1 mg/mL)	2 μL	2 μL	2 μL	2 μL	2 μL	2 μL
*Bam*HI	1 μL	1 μL	1 μL	1 μL	1 μL	1 μL
*Nde*I	1 μL	1 μL	1 μL	1 μL	1 μL	1 μL
Water	4 μL	4 μL	4 μL	4 μL	4 μL	4 μL
Final volume	20 μL	20 μL	20 μL	20 μL	20 μL	20 μL

2. For *Nde*I digestions, put in 0.5 μL first and let the reactions go for 30 min before putting in the second half of *Nde*I.

3. Digest overnight at 37°C.

Day 7

Today you run agarose gels to show that the plasmids were cut correctly with the restriction enzymes. This is followed by excision of the correct bands from the gel.

Running Agarose Gel Electrophoresis

1. Make two 40-mL gels of 1% agarose in 1× TAE buffer (40 mM Tris-acetate, 1 mM EDTA, pH 8.0). Heat in a microwave until dissolved (about 30–45 s).

2. Pour the agarose gels and insert the combs. After the gels solidify, assemble the apparatus and cover the gels with TAE buffer.

3. Mix your overnight digestions with bromophenol blue/xylene cyanol tracking dye to the correct concentration (that is, 20 μL of digestion plus 4 μL of 6× tracking dye).

4. Make up a 10-μL sample of undigested plasmid of each type and mix with tracking dye.

5. Acquire two DNA ladder samples. One will be *Hin*dIII fragments of λ DNA. The other will be a low–molecular weight marker that starts at 3 kb and decreases. Add tracking dye to these unless they have already had this done (that is, they are already blue).

6. Heat all samples mixed with tracking dye at 65°C for 2 min and then put on ice.

7. Load all samples, splitting them into two wells when necessary. Electrophorese at 100 V until the tracking dye has traveled three fourths of the way to the end.

8. Place the gel into a container with 30–50 mL of buffer, add 10 μL of 10 mg/mL ethidium bromide, and let it soak into the gel for 10 min. This may also be done by adding 10 μL to each chamber of the electrophoresis unit when the gel is half done.

9. Use a UV light box to observe the bands. Keep the UV light on for very short times.

10. Excise the LDH bands (1 kb) and the pET 15b bands (5.7 kb). The best way is to quickly slice above each band while the UV light is on and then continue to remove slices after you have turned off the light. You want to get all the DNA but not extra gel in the slice.

11. Place the gel slices for the LDH into one microfuge tube and the slices for the pET 15b into another. Label and freeze for the next lab time.

Name _____   Section _____

Lab partner(s) _____   Date _____

Analysis of Results

Experiment 12: **Cloning and Expression of Barracuda LDH-A: Part I**

Data

Sketch or provide a picture of your agarose gel showing the DNA ladders and the digested and undigested plasmids.

Questions

1. Why is it necessary to make sure the LB–agar solution is cool enough before adding the ampicillin? What would your plates look like if you added the ampicillin when the solution was too hot?

2. How many bands did you expect to see in your digestion lanes on the agarose gel? How many did you see? Explain any discrepancies.

3. Why do you heat the DNA samples plus tracking dye at 65°C for 2 min?

4. Why is pET 15b being used for an expression vector?

5. Why is the word *gently* emphasized so often in the procedure for making competent cells? What is the procedure doing to the cells?

6. Given that pET 15b can be bought, why is pET 15b that already has a DNA insert of rat kidney glutaminase (rKGA) DNA being used?

7. In step 15 on day 3 when doing transformations, why are two different volumes of transformed cells plated on LB/amp?

Experimental Procedures: Part II

This part of the experiment takes you through the ligation of LDH DNA to pET 15b and the transformation of competent cells.

Day 8

Today you extract the DNA out of the gel slices saved from last time. **Remember that the gel slices have EtBr in them, so you need to wear gloves.** You also start cell cultures for the next lab.

Purifying Excised DNA with the Bio 101 Gene Clean Gel Extraction Kit

1. To the tubes with the gel slices, add three volumes of the NaI solution (for example, if the slice weighs 0.3 g, add 300 μL of NaI).

2. Incubate for 5 min at 45–55°C.

3. Add 5 μL of glassmilk solution that has been vortexed for 30 s.

4. Mix the tubes and incubate on ice for 5 min.

5. Spin the tubes for 5 s at full speed.

6. Remove the supernatants and save. The supernatant should not have any DNA in it, but it is saved just in case something goes wrong.

7. Resuspend the pellet in 700 μL of new wash solution. Spin for 5 s and remove the supernatant.

8. Repeat step 7 twice more to fully wash the pellet.

9. Resuspend with 5 μL of TE buffer (10 mM Tris, 1 mM EDTA, pH 8.0) or water.

10. Spin for 30 s.

11. Remove the supernatant and save.

12. Resuspend the pellet in 5 μL of TE or water, spin, and remove the supernatant. Combine with the supernatant from step 11. These combined supernatants are your DNA samples.

Starting Cell Cultures for Ligation and Transformations

1. Start an overnight culture of cells (DH5α, JM109, or BL21), using 2 mL of medium C and 50 μL of saved culture from the first day.

2. Place the tubes in a shaking incubator at 37°C overnight.

Day 9

This is the most critical day. Today you ligate the LDH DNA to the pET 15b vector. To do this, you use varying combinations of the vector and insert. One of the combinations will hopefully work well. You also run a control that has no insert to show that the pET 15b cannot reclose upon itself.

1. Put 1.5 mL medium C into each of six clear-capped tubes and warm to 37°C.

2. Add 150 µL overnight culture to each tube.

3. Place the tubes in a shaking incubator at 37°C for 20 min.

4. Put six LB/amp plates in a 37°C incubator and save for step 18.

5. Set up ligation reactions according to the following table (the LDH and the pET 15b are from the samples from the Gene Clean kit).

| | Tube | | | |
Reagent	1	2	3	4
pET 15b	6 µL	3 µL	6 µL	6 µL
LDH insert	3 µL	6 µL	6 µL	
5× ligase buffer	4 µL	4 µL	4 µL	4 µL
Water	6 µL	6 µL	3 µL	9 µL
DNA ligase	1 µL	1 µL	1 µL	1 µL
Total	20 µL	20 µL	20 µL	20 µL

6. Vortex the tubes and spin down in a microfuge.

7. Incubate for 1 h on ice.

8. Spin the tubes from step 3 for 1 min at 4°C and discard the supernatants.

9. Mix T-solution A with T-solution B in a 1:1 ratio to give the final T-solution for the following steps.

10. Add 300 µL of T-solution to the tubes and *gently* resuspend.

11. Keep on ice for 5 min.

12. Spin again at 4°C for 1 min. Discard the supernatants.

13. Add 120 μL of T-solution to the tubes.

14. Keep on ice for 5 min.

15. Aliquot 5 μL of each reaction from the preceding table to clean microfuge tubes. Also aliquot 5 μL of a control plasmid to a separate tube and have a tube for a no-DNA control. *Freeze the leftover ligation mixes and label carefully.* You will need these later.

16. Add 50 μL of the T-cells to each of the six tubes (ligations, control plasmid, and no DNA).

17. Keep on ice for 5 min.

18. Plate the entire mix on six LB/amp plates.

19. Incubate overnight at 37°C.

Day 10

Sometime in the afternoon, recover your plates from the incubator and put them in a refrigerator at 4°C.

Day 11

Preparing Overnight Cultures

1. Recover your plates from the refrigerator.

2. Choose five colonies from your ligation plates. If you have colonies on more than one ligation plate, choose some from each. For each one, add a loopful to 5 mL of L-broth in sterile Falcon tubes. Add 5 μL of 100 mg/mL ampicillin. Make sure that the caps are not on tight. Tape the caps loosely so that they allow oxygen in.

3. Place in a 37°C shaking incubator overnight.

Day 12

Today you do plasmid preps to determine if you had any pET 15b/LDH recombinants, and you start restriction enzyme digestions.

Setting Up Plasmid Preps (Qiagen)

1. Spin down 1.7 mL of each of your pET 15b/LDH overnight cultures.

2. Resuspend the bacterial pellets in 250 μL of buffer P1, which should have had RNase A added already.

3. Add 250 μL of buffer P2, mix *gently* by inversion four to six times or until the solution becomes viscous. Incubate at room temperature for 5 min. Do not let this step go for longer than 5 min.

4. Add 350 μL of chilled buffer N3, mix immediately and *gently* by inversion four to six times, and incubate on ice for 5 min.

5. Mix the samples again and centrifuge at maximum speed in a microfuge for 10 min. Remove the supernatants promptly. The supernatants should be clear.

6. Apply the supernatants from step 5 to the spin columns.

7. Centrifuge for 1 min. Discard the flow-through.

8. Add 0.5 mL of buffer PB to the columns and centrifuge for 1 min. Discard the flow-through.

9. Add 0.75 mL of buffer PE to the spin columns and centrifuge for 1 min. Discard the flow-through.

10. Centrifuge the columns for 1 min to remove the last traces of wash buffer.

11. Place the spin columns in clean 1.5-mL microfuge tubes and add 50 μL of buffer EB to the center of the columns. Let stand for 1 min and then spin for 1 min.

12. The solutions that flowed through into the microfuge tube are your purified plasmids.

13. As before, the Wizard prep or the quick-and-dirty prep could be used instead of this Qiagen procedure.

Plasmid Digests

In this part, you cut the Qiagen purified plasmids with *Nde*I and *Bam*HI. Be sure to save all your plasmid preps because they will be needed for backup if future parts of this experiment fail.

You should have five samples to use. For each one, set up a digestion according to the following protocol:

1. For the *Nde*I digestion, put in 0.5 μL first and let the reaction go for 30 min. Then put in the second half of *Nde*I.

2. Digest overnight at 37°C.

3. Remember to save all plasmid preps that you did not digest and label them well so that you can later find your best sample.

pET 15b/LDH	10 μL
10× *Bam*HI buffer	2 μL
BSA 1 mg/mL	2 μL
*Bam*HI	1 μL
*Nde*I	1 μL
Water	4 μL
Final volume	20 μL

Day 13

Today you run an agarose gel to show that you had plasmid with inserts.

Running Agarose Gel Electrophoresis

1. Make a 40-mL gel of 1% agarose in 1× TAE buffer (40 mM Tris-acetate, 1 mM EDTA, pH 8.0). Heat in a microwave until dissolved (about 30–45 s).

2. Pour the agarose gel as before and insert the comb. Assemble the apparatus and cover the gel with TAE buffer.

3. Mix your overnight digestions with tracking dye to the correct concentration (20 µL of digestion plus 4 µL of 6× tracking dye).

4. Acquire two DNA ladder samples. One will be *Hin*dIII fragments of λ DNA. The other will be a low–molecular weight marker that starts at 3 kb and decreases. Add tracking dye to these unless they have already had this done (that is, they are already blue).

5. Heat all samples mixed with tracking dye at 65°C for 2 min and then put on ice.

6. Load 15 µL of the samples and electrophorese at 100 V until the tracking dye has traveled three fourths of the way to the end.

7. Place the gel into a container with 30–50 mL of buffer and add 10 µL of 10 mg/mL ethidium bromide and let it soak into the gel for 10 min. This may also be done by adding 10 µL to each chamber of the electrophoresis unit when the gel is half done.

8. Use a UV light box to observe the bands. Keep the UV light on for very short times.

Preparing LB/Amp Plates

Depending on how many plates were made on the first day and subsequently used, you might need to make some new ones for the rest of the experiment. If so, follow the same procedures as before.

TIP 12.5 **Caution:** Wear gloves, goggles, and a face shield when looking at the light box because this UV light will burn your skin quickly.

Setting Up Expression Cell Cultures

1. Inoculate three tubes of 2 mL of TransformAid C-medium with 20 µL of BL21 DE3 cells from frozen, commercial competent cells.

2. Incubate the cultures overnight at 37°C.

Analysis of Results

Experiment 12: **Cloning and Expression of Barracuda LDH-A: Part II**

Data

1. Describe the results of your transformation plates:

   pET 15/LDH ligation:

   pET 15 without LDH:

   Control plasmid:

   No-DNA control:

2. Provide a picture or sketch of your agarose gel showing your plasmid digests.

Questions

1. Which plate had the most colonies on it? Is this what you expected? Why?

2. What does it mean if pET 15b without LDH has colonies?

3. What does it mean if the no-DNA control has colonies?

4. What were the results of your digestions? What bands showed up on your agarose gels?

Experimental Procedures: Part III

In this part, you transform the expression cell line with the pET 15/LDH plasmid and express the LDH fusion protein. As a final step, the LDH fusion protein is purified using a nickel affinity column.

Day 14

Transforming BL21 DE3 Cells

1. Put 1.5 mL of C-medium into each of three microfuge tubes and warm to 37°C.
2. Add 150 μL of overnight cultures to the tubes.
3. Place in a 37°C shaking incubator for 20 min.
4. Place the tubes on ice and spin each tube at 4°C for 1 min. Discard the supernatant.
5. Mix T-solution A with T-solution B in a 1:1 ratio to give the final T-solution for the following steps.
6. Add 300 μL of T-solution to the tubes and *gently* resuspend.
7. Keep on ice for 5 min.
8. Spin again at 4°C for 1 min. Discard the supernatant.
9. Add 120 μL of T-solution to the tubes.
10. Keep on ice for 5 min.
11. Aliquot 10 μL of pET 15/LDH to a 1.5-mL microfuge tube.
12. Aliquot 10 μL of a control plasmid to a 1.5-mL microfuge tube.
13. Put both tubes on ice for 5 min along with a third control, an empty microfuge tube.
14. Add 50 μL of T-cells to each DNA tube and the rest to the control tube.
15. Put all tubes on ice for 5 min.
16. Plate the DNA and control cell mixes on LB/amp plates.
17. For the control plate, plate the whole mix on one plate.
18. For the plasmid plates, do two plates for each plasmid. One should be 10 μL and the other 40 μL.
19. Incubate the five plates overnight at 37°C.

Preparing SDS-PAGE

You will eventually be using SDS-PAGE to help verify that a foreign protein was induced. Today is a convenient time to make these gels.

1. Make a 12% SDS separating gel and a 3–5% stacking gel as you did in Experiment 9c.

2. Store the gels (with the combs still in) at 4°C.

Day 15

Today is a short day to begin cultures from your successful plates.

Preparing Overnight Cultures

1. Sometime in the afternoon of the day following transformation, recover your plates.

2. Choose three colonies from your pET plus LDH plates and add a loopful to 10 mL of L-broth in sterile Falcon tubes. Add 10 μL of 100 mg/mL amp.

3. Place in a 37°C shaking incubator overnight.

Day 16

The following procedure begins early in the morning to generate cells that express LDH. Overnight cultures of BL21 DE3 carrying the pET 15/LDH are scaled up to 2 L. They are grown at 37°C for 2 h until their OD_{600} is about 0.4. They are then split into two batches. One batch of 500 mL is left uninduced. The other batch of 1500 mL is induced with IPTG at a final concentration of 1 mM. These are then transferred to a 30°C incubator and allowed to express for 3 h.

Harvesting Cells

1. You should have 250 mL of induced and uninduced cells in pre-weighed bottles.

2. Spin the cells at 5000 × g for 20 min.

3. Discard the supernatant.

4. Weigh the bottles to determine the weight of the cell precipitate.

5. Add 10 volumes of binding buffer (20 mM sodium phosphate, 0.5 M NaCl, pH 7.4). Resuspend the cells by pipetting up and down until the cells are completely suspended.

6. Sonicate the cells four times for 30 s each, with 30 s rest on ice in-between to avoid overheating.

7. Spin the sonicated cells at 10,000 × g for 15 min.

8. Save the supernatant. To be safe, also save the pellet.

9. Save 500 μL of the induced and uninduced cell supernatant to use for later enzyme assays.

10. Keep everything on ice when not in use.

Preparing and Loading Columns

In this step, you purify the induced LDH with a 5-mL nickel affinity column called HisTrap. These procedures vary greatly depending on what type of column system you have. The flow rate through the column should be 1 mL/min.

1. Wash the column with at least 5 mL of water.

2. Equilibrate the column with 10 mL of binding buffer.

3. Load the induced sample through the column. Near the end of the sample-loading process, collect 1 mL of the solution flowing through the column (flow-through). These are the proteins that are not binding to the nickel.

4. When the last of the sample is loaded, switch back to the binding buffer.

5. If your system has a UV detector, wash with binding buffer until the absorbance is back down close to 0. If you do not have a UV monitor, wash with 25 mL of binding buffer.

Eluting LDH

1. Switch the buffer flowing through the column to the elution buffer (20 mM sodium phosphate, 0.5 M NaCl, 500 mM imidazole, pH 7.4).

2. Elute the LDH at 1 mL/min, collecting 1-mL fractions. If you have a UV monitor, note when the absorbance at 280 nm increases rapidly. This will be when the imidazole is going through the detector.

3. Collect 15 to 20 fractions. It should take 4–6 mL of elution buffer to elute the LDH once the buffer begins exiting the column.

4. Continue washing the column with elution buffer to be sure that bound proteins are removed. When done eluting, wash the column with distilled water.

Enzyme Assays

1. Assay your fractions for LDH using the procedures from Experiment 4a.

2. Assay all of the following:

    • Uninduced cell supernatant

    • Induced cell supernatant

- Flow-through from the column wash
- Column fractions

Day 17

Running SDS-PAGE

The last day is devoted to running SDS-PAGE to show the increase in LDH between the induced and uninduced cells. Run the following samples on the gel, using the same procedures as Experiment 9c.

- Dalton VII marker
- Glyceraldehyde-3-phosphate marker (this has the same molecular weight as the LDH monomer for comparison)
- Uninduced cell supernatant
- Induced cell supernatant
- HisTrap column flow-through
- Five column fractions centered about the fraction that had the most activity

Name _____

Section _____

Lab partner(s) _____

Date _____

Analysis of Results

Experiment 12: **Cloning and Expression of Barracuda LDH-A: Part III**

Data

1. Draw a sketch of your SDS-PAGE gel.

2. Fill in the following table for the enzyme fractions.

Fraction	LDH (units/mL)
Uninduced supernatant	
Induced supernatant	
Column flow-through	
Fraction no. _____	
Fraction no. _____	
Fraction no. _____	
Fraction no. _____	
Fraction no. _____	
Fraction no. _____	
Fraction no. _____	
Fraction no. _____	
Fraction no. _____	
Fraction no. _____	

Questions

1. What do the results show concerning the induced and uninduced samples?

2. What type of purification did you see with the LDH fusion protein?

Additional Problem Set

1. Explain and diagram how blue/white screening works. How does this compare to the earlier selections using ampicillin- and tetracycline-resistance markers?

2. What are the general requirements for a plasmid used in cloning?

3. What are the general requirements for a plasmid used in expression of foreign proteins?

4. Outline the procedures used to produce human erythropoietin in bacteria. What might such a compound be used for?

5. The gene for β-globin is a split gene, which means it contains introns. How does that affect a plan to clone human β-globin in bacteria?

6. If you want to know the sequence of a protein, what are the advantages and disadvantages of sequencing the protein directly versus sequencing the DNA for the protein?

7. A vector has a polylinker (MCS) containing restriction sites in the following order: *Hind*III, *Sac*I, *Xho*I, *Bgl*II, *Xba*I, and *Cla*I.

   a. Give a possible nucleotide sequence for the polylinker.

   b. The vector is digested with *Hind*III and *Cla*I. A DNA segment contains a *Hind*III restriction site 650 bases upstream from a *Cla*I site. This DNA segment is digested with *Hind*III and *Cla*I, and the resulting *Hind*III–*Cla*I fragment is directionally cloned into the digested vector. Give the nucleotide sequence at each end of the vector and the insert and show that the insert can be cloned into the vector in only one orientation.

Webconnections

For a list of web sites related to the material covered in this chapter, go to **Webconnections** at the *Experiments in Biochemistry* site on the Brooks/Cole Publishing web site. You can access this page at *http://www.brookscole.com* and follow the links from the chemistry page.

References and Further Reading

Adolph, K. W. *Advanced Techniques in Chromosome Research.* New York: Marcel Dekker, 1991.

Ausubel, F. M., R. Brent, R. Kingston, et al. *Current Protocols in Molecular Biology.* New York: Wiley, 1987.

Beugelsdijk, J. *Automation Technologies for Genome Characterization.* New York: Wiley Interscience, 1997.

Boyer, R. F. *Modern Experimental Biochemistry.* Menlo Park, CA: Addison-Wesley, 1993.

Campbell, M. K. *Biochemistry.* Philadelphia: Harcourt Brace College, 1998.

Crawford, D. L., H. R. Constantino, and D. A. Powers. "Lactate Dehydrogenase-B cDNA from the Teleost *Fundulus heteroclitus:* Evolutionary Implications." *Molecular Biology of Evolution* 6, no. 4 (1989).

Garrett, R. H., and C. M. Grisham. *Biochemistry.* Philadelphia: Saunders, 1995.

Holland, L. Z., M. McFall-Ngai, and G. N. Somero. "Evolution of Lactate Dehydrogenase-A Homologs of Barracuda Fishes from Different Thermal Environments." *Biochemistry* 36 (1997).

Huang, D., C. J. Hubbard, and R. A. Jungmann. "Lactate Dehydrogenase-A Subunit Messenger RNA Stability Is Synergistically Regulated via the Protein Kinase A and C Signal Transduction Pathways." *Molecular Endocrinology* 9, no. 8 (1995).

Maekawa, M., K. Sudo, S. S. Li, and T. Kanno. "Genotypic Analysis of Families with Lactate Dehydrogenase A (M) Deficiency by Selective DNA Amplification." *Human Genetics* 88, no. 1 (1991).

Miyajima, H., T. Shimizu, and E. Kaneko. "Gene Expression in Lactate Dehydrogenase-A Subunit Deficiency." *Rinsho Shinkeigaku* 32, no. 10 (1992).

Promega Corporation. *Technical Manual for pET-5 Expression Vectors.* San Luis Obispo, 1995.

Quattro, J. M., H. A. Woods, and D. A. Powers. "Sequence Analysis of Teleost Retina-Specific Lactate Dehydrogenase C: Evolutionary Implications for the Vertebrate Lactate Dehydrogenase Gene Family." *Proceedings of the National Academy of Sciences* 90, no. 1 (1993).

Wyckoff, H. A., J. Chow, T. R. Whitehead, and M. A. Cotta. "Cloning, Sequence, and Expression of the L-(+) Lactate Dehydrogenase of *Streptococcus bovis.*" *Current Microbiology* 34 (1997).

Zhou, W., and E. Goldberg, "A Dual-Function Palindromic Sequence Regulates Testis-Specific Transcription of the Mouse Lactate Dehydrogenase C Gene in Vitro." *Biology of Reproduction* 54, no. 1 (1996).

Chapter 13

Polymerase Chain Reaction

TOPICS

Introduction

In this final chapter, we deal with the exciting and popular technique of polymerase chain reaction, a method for amplifying DNA in microgram amounts from nanograms or less of starting material.

13.1 Amplification of DNA

In Chapter 12, we discussed how cloning could be used to increase the amount of a specific gene. With cloning, we let the host cell's DNA replication system amplify the target DNA. In this chapter, we discuss another method used to amplify DNA without having to go through the cloning process: the **polymerase chain reaction** (PCR).

PCR technology makes it possible to amplify a gene without cloning it from the rest of its DNA. This target gene can be in a plasmid, a virus, or a chromosome; it can also be a piece of naked DNA. PCR is so sensitive that it can produce over 1 µg of specific target DNA from less than 50 ng in a few hours.

PCR is an automated procedure carried out in a machine called a **thermocycler** that controls the time and temperature of the amplification reactions. The key to the process was the discovery of a heat-stable DNA polymerase (*Taq*) from the bacteria *Thermus aquaticus*. This enzyme allows the reactions to proceed at high temperatures without denaturing the key enzyme that replicates the DNA. Other components of the reaction include the target DNA that you want to amplify, the full set of deoxynucleoside triphosphates (dNTPs), and specific primers that are complementary to the DNA in or near the target DNA.

13.2 *Taq* Polymerase

Many enzymes are involved when an organism replicates its DNA, and the reaction occurs at biological temperatures (20–40°C). Some enzymes are involved in the synthesis and subsequent proofreading and repair of the new

DNA. Other enzymes are involved in the unwinding of the DNA double helix so that replication can proceed. The complexity of this process, and the requirement for ambient temperatures so that enzymes can work, make it impossible to automate a DNA replication outside the organism. If we want to produce DNA chemically, we have to find a way to separate DNA strands first. This requires a temperature that inactivates the average DNA polymerase used to replicate the single strand.

In 1993, Dr. Kary Mullis won the Nobel Prize in science for his PCR procedure. He used the DNA polymerase from *T. aquaticus*, known now as *Taq* polymerase. The bacteria were found living in undersea vents at temperatures near 100°C. Any enzymes from these bacteria have to work at high temperatures. Using *Taq* polymerase allowed the automation of PCR. A thermocycler can raise the temperature high enough to separate strands of DNA without killing the enzyme. Next the thermocycler reduces the temperature, allowing primers to anneal. The temperature is then raised again to the one optimal for the polymerase to work. The polymerase copies the single strands, starting at the primer and going from 5′ to 3′ as all DNA synthesis does. Once enough time has elapsed to replicate the required DNA sequence, the thermocycler raises the temperature again to separate DNA strands. Each round of replication doubles the amount of DNA. It takes about 1 minute to replicate 1 kilobase (kb) of DNA. So, with this process, if the target DNA sequence is 1 kb, the sequence can be doubled every couple of minutes. At that rate, a huge amplification occurs in a few hours. Figure 13.1 demonstrates the basics of this reaction.

The shaded DNA in the middle is the target sequence. The reaction begins with the separation of strands (step 1). A common way to do this is to heat samples to 95°C for 30 seconds. In step 2, the reactions are cooled to 60°C for 30 seconds; this allows primers to anneal. Primers are shown as the short DNA sequence that is attached just outside the target sequence. These primers are always present in excess so that, as soon as DNA is separated, primers are ready to bind. In step 3, samples are heated to 72°C to allow the *Taq* polymerase to extend the DNA, starting with primers and heading toward the 3′ end. That is the end of one cycle. This cycle is repeated as often as necessary to obtain the desired amount of target DNA. It is important to note that the amount of both primer and dNTPS actually controls the maximum possible PCR product. Once either runs out, the reaction stops.

One problem associated with PCR is that, by using the artificial system of replication, the proofreading and repair functions that many organisms have as part of their DNA replication are lost. *Taq* polymerase makes a high number of errors when it synthesizes DNA. If one of these errors happens early, that mistake is amplified many times, and the product will be different from the starting target DNA. DNA sequencing is often necessary to verify that the PCR product is correct. Other heat-stabile DNA polymerases have since been discovered, and some additional enzymes that confer a proofreading capability, such as **Vent Polymerase,** are also used.

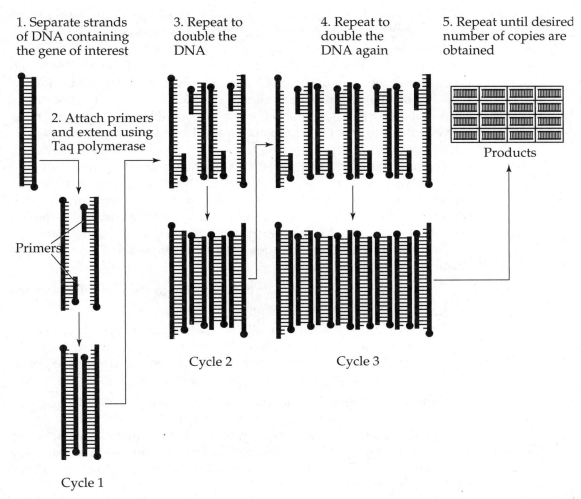

1. Separate strands of DNA containing the gene of interest

2. Attach primers and extend using Taq polymerase

Primers

3. Repeat to double the DNA

4. Repeat to double the DNA again

5. Repeat until desired number of copies are obtained

Products

Cycle 2

Cycle 3

Cycle 1

FIGURE 13.1 *The polymerase chain reaction*

13.3 Primers

Designing primers is the most important part of a successful PCR. Figure 13.1 shows the primers binding outside the target DNA; this is one way to do it. Another way is to have primers be part of the target DNA sequence. In either case, the target DNA is replicated.

Primers are 18 to 40 bases long. If a primer is too short, it will not bind specifically enough to the desired region of DNA. If it is too long, it becomes expensive and the annealing time becomes longer. Each strand of DNA must have its own primer, called the forward and reverse primers. These will have different sequences because they are designed to bind to different pieces of DNA. It is important to make sure that primers won't anneal to one another or to similar DNA that is not part of the target gene.

Custom primers are those that you specify and order for your particular PCR needs. For example, if we want to do PCR on the barracuda

LDH-A gene and we know its sequence, we can specify the exact bases that are complementary to the first 25 bases at the end of each strand.

Standard primers are those that are used with common vectors. If your gene of interest is in pBlueScript, for example and the gene is cloned into the multiple cloning site (MCS), you can take advantage of the known sequence of this vector on each side of the MCS. Primers designed to the DNA outside the MCS bracket your target DNA so that everything between these primers is replicated. Figure 13.2 shows a map of how this might work.

Outside the MCS, you can find known sequences based on the T3 RNA polymerase promoter and the T7 RNA polymerase promoter. Therefore, if you order the T7/T3 primer set, you can replicate via PCR whatever target DNA you have inserted into the MCS. Another common primer set is the M13-20/Reverse primers, also shown in Figure 13.2.

Primers must be chosen or designed with great care. These are some characteristics of good primers:

1. Primers must be long enough to be specific to the target DNA or the DNA just surrounding the target.

2. Primers must be short enough to meet your budgetary concerns. Clearly, if you can make a 1000-base primer, it will be very specific, but you probably have to synthesize the entire gene. If you can afford to do that and it were possible, you don't need to use PCR.

3. Forward and reverse primers should have similar G + C contents. Remember that the melting temperature of DNA is dependent on the G + C content. Thermocycler programs are optimized for annealing temperatures and times. If forward and reverse primers have radically different G + C contents, then the same program will not work well for both primers.

4. Primers should have low complementarity to each other. You want primers to bind to the target DNA or that which surrounds it. If your forward primer is 5'-AAATTTAAATTT-3' and your reverse primer is 5'-TTTAAATTTAAA-3', then your two primers want to bind to each other instead of the target DNA. This removes primers from the reaction, and you will get no product.

5. Primers should have minimal secondary structure. Primers shown in characteristic 4 are unsatisfactory for another reason. Each easily binds to itself, making a hairpin loop:

5'-AAATTT
3'-TTTAAA

This also eliminates the primer from binding with the target, so your reactions will not continue.

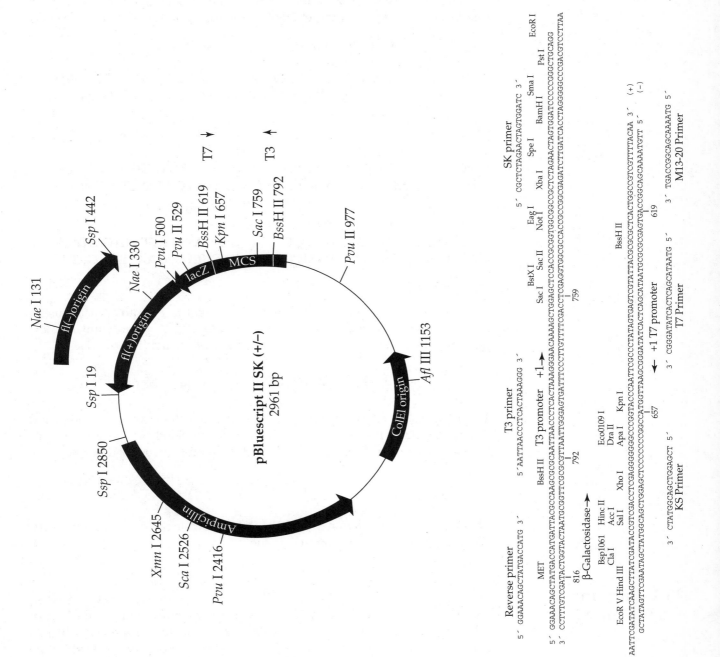

FIGURE 13.2 *pBluescript and partial sequence*

PRACTICE SESSION 13.1

If you have the LDH gene cloned in pBlueScript as shown below, what primers can you use to amplify the gene by PCR? The LDH gene is inserted into the vector with *Bam*HI and *Hin*dIII. The vector sequence is shown in boldfaced italic type; the target DNA sequence is shown in boldfaced type.

```
        BamHI           Start  S     T     K
          ↓               ↓    ↓     ↓     ↓
5'-ACT AGT GGA TCC  ATG TCC  ACC AAG GAG AAG CTC  ATC
3'-TGA TCA CCT AGG  TAC AGG TGG TTC  CTC TTC  GAG TAG
                ↑
```

```
                         L     T     L    Stop
                         ↓     ↓     ↓     ↓
GAC CAC GTG ATG // more LDH // CTC ACC CTG TGA CCT
CTG GTG CAC TAC // more LDH // GAT TGG GAC ACT GGA
```

```
                             HindIII
                                ↓
CTG ATT TCT CCA GTC CGC CTT  GAA AAG CTT  TAG CTA-3'
GAC TAA AGA GGT CAG GCG GGA CTT  TTC GAA ATC GAT-5'
                                          ↑
```

Let's assume that we want our PCR product to go into a vector using the same two restriction sites, *Bam*HI and *Hin*dIII. We then want primers that include these sequences. A possible forward primer is:

5'-GGA TCC ATG TCC ACC AAG GAG-3'

This matches the restriction site at the 5' end and includes the start codon and four amino acids. If this primer proves to be insufficiently specific, increase its length by adding more bases at the 3' end. A possible reverse primer is

5'-AAG CTT TTC AGG GCG GAC TGG-3'

This retains the restriction site at the 3' end and includes several noncoding bases that are part of the LDH gene. ●

13.4 Changing Restriction Sites

One of the most powerful uses for PCR is changing the restriction sites at the ends of the target DNA. Let's assume that we cut the LDH gene out of pBlueScript as in Practice Session 13.1. If we want to put the gene into an expression vector but the expression vector does not have a *Bam*HI site, can we use it? The answer is yes. With PCR we can change the sequence and put in a different restriction site. A common restriction site used with expression vectors is *Nde*I, which recognizes the sequence CATATG. This is common with expression vectors because it makes use of the start codon. Every gene has to have an ATG sequence, and using *Nde*I eliminates upstream noncoding regions.

ESSENTIAL INFORMATION

Polymerase chain reaction (PCR) is a very powerful technique that allows the rapid and specific production of target DNA. It is not based on cloning technology, and no organisms need to be grown to produce the DNA. The target DNA can be free or in a suitable vector. PCR is possible due to the discovery of heat-stable forms of DNA polymerase. This allows an automated reaction sequence of dissociation of double-stranded DNA, binding of primers, and extension with the polymerase. The design of primers is the most critical part of PCR. The primer must bind at the right place on or near the target DNA, must not bind to itself, and must have the proper length to meet specificity and cost considerations. Using PCR technology, individual bases can be changed from the orginal target DNA. This might be done to change the restriction sites at the end of a gene being inserted into another vector, or it might be done to mutate the actual DNA sequence of the gene.

If we want to clone the LDH gene into an expression vector using an *Nde*I site, just change the forward primer to include the *Nde*I sequence. PCR then will produce a product with the proper ends. Instead of the primers used previously, we use

CAT ATG TCC AC AAG GAG

In the first cycle, the CAT has nothing to bind to because the 3' strand is not complementary at that position:

CAT -ATG TCC ACC AAG GAG
TGA TCA CCT AGG TAC AGG TGG TTC CTC

However, enough of the primer is complementary to make it bind, and the CAT would just "go along for the ride" on the first cycle. After the first cycle, the replication of the 5' strand would create the complement to the CAT, such that the 3' strand would start as

3'-GTA TAC AGG TGG TTC CTC

Now the PCR product begins with the *Nde*I site.

This method is also used to change individual bases within the gene product, in a powerful technique called **site-directed mutagenesis.**

13.5 Why Is This Important?

PCR technology is quickly becoming the most important single biotechnology technique that you can learn. It is amazingly simple once you have acquired the correct primers and learn how to program the thermocycler. With PCR, minuscule samples of DNA can be amplified to give microgram quantities for analysis. In many labs, PCR is replacing classical cloning procedures for scale-ups of DNA. With PCR, individual bases can be changed in the important technique of site-directed mutagenesis. These bases might be changed to allow the DNA to be inserted more easily into a vector of choice. They might be changed to study the resulting structure or function differences of the translated protein. Forensics uses PCR to obtain enough DNA from a crime scene to eventually identify or exclude a potential criminal.

PCR can also be used to study transcription and to quantitate different levels of DNA or RNA. Reverse transcriptase PCR (RT-PCR) involves using reverse transcriptase to make cDNA from mRNA. The DNA can then be amplified using PCR. Nucleic acids can be quantitated using Q-PCR. Fluorescent dyes are used to bind to the amplified PCR products. By calculating the number of product molecules, the original number of target molecules can be determined.

Experiment 13

Polymerase Chain Reaction of Barracuda LDH-A

In this experiment, you will use another technique to amplify the DNA from barracuda LDH-A—namely, PCR.

Prelab Questions

1. How is PCR different from cloning?

2. What is meant by a *forward* or a *reverse primer*?

3. What are characteristics of a good primer?

4. What are our primers complementary to?

5. What is the purpose for each step in the thermocycler routine?

Objective

Upon successful completion of this lab, you will be able to

- Use PCR to amplify the gene for LDH-A from the barracuda, *Sphyraena lucasana*.

Experimental Procedures

Standard Procedure

1. Set up three reactions as shown in the following table. These include a negative control with no DNA template added, a positive control using pUC18 to verify that the PCR reaction worked with known primers, and a reaction with the PCRScript/LDH plasmid. It is best to use a cocktail for much of this.

	No DNA Control (μL)	pUC18 DNA (μL)	PCRScript (μL)
DNA template	0	1	1
Primer 1	5	0	5
Primer 2	5	0	5
Control primer 1	0	5	0
Control primer 2	0	5	0
Taq polymerase	0.5	0.5	0.5
dNTP	16	16	16
10$\times$ PCR buffer	10	10	10
H$_2$O	63.5	62.5	62.5

The forward primer is

5'-C CAT ATG TCC ACC AAG GAG AAG CTC ATC GAC CAC GTG ATG

The reverse primer is

5'-TTC AGG GCG GAC TGG AGA AAT CAG AGG

The control primers are standard M13–20 forward and reverse.

Note: Volumes are adjusted to give 100 pmol of each primer and 1–100 pmol of template, depending on the source. We use 1 pmol. The dNTP stock is 2 mM, and the *Taq* stock is 5 U/μL.

2. The program used for the thermocycler cycles is:

   a. Start at 95°C for 5 min (denaturation)

   b. 30 cycles of the following:

      i. 95°C for 1 min (denaturation)

      ii. 55°C for 1 min (annealing primers)

      iii. 72°C for 1 min (extension of primers)

   c. 4°C overnight

PCR Bead Method

Another way to set up PCR is to use PCR beads, which have all enzymes, buffers, and dNTPs necessary for the reaction. Just add your primers and template.

1. Set up the protocol shown in the following table for reactions, using PCR beads.

	No DNA Control (μL)	pUC18DNA (μL)	PCRScript (μL)
DNA template	0	1	1
LDH primer 1	5	0	5
LDH primer 2	5	0	5
M13 primer 1	0	5	0
M13 primer 2	0	5	0
H$_2$O	15	14	14

2. If the thermocycler does not have a hot lid, add 50 μL of mineral oil to the tubes and start the same program as shown in step 1.

Gel Electrophoresis of PCR Products

1. Prepare a 1.25% agarose gel in TAE buffer. This is similar to the gels we made in Experiment 12, but the agarose concentration is a little higher.

2. Remove 10 µL of the PCR reactions (from underneath the mineral oil, if necessary) and add 2 µL of 6× loading buffer. Heat at 65°C for 5 min and then place on ice.

3. Run lanes with λ DNA/*Hin*dIII fragments and low molecular weight DNA markers as in Experiment 12 (also heated to 65°C first).

4. Load the gel and electrophorese at 80–100 V.

5. Soak the gel in 50 mL of buffer or water with 10 µL of 10 mg/mL ethidium bromide for 15 min. Photograph on a UV lamp.

Analysis of Results

Experiment 13: **Polymerase Chain Reaction of Barracuda LDH-A**

Data

1. Sketch or provide a picture of your agarose gel, showing the MW markers, control reaction, and LDH bands.

2. What are the sizes of the fragments generated by PCR?

Questions

1. What is the purpose of using the pUC18 plasmid and M13 primers?

2. What is the purpose of having the "No DNA" lane? What does it mean if this lane has bands?

3. Why is it impossible to have automated PCR without a heat-stable form of DNA polymerase?

4. If you have no heat-stable form of DNA polymerase but want to do PCR, what must you do each cycle?

5. Using primers listed in the procedures, can you take the PCR product and ligate into the pET15 vector used in Experiment 12? Why or why not?

6. What is the purpose of the mineral oil used for PCR when the thermo-cycler does not have a hot lid?

7. How do you change the reverse primer if you want to change the *Eco*RI site for a *Bam*HI site?

8. Describe what happens if you let the extension phase of the PCR go too long with each cycle. Will your final product be longer? Why or why not?

Additional Problem Set

1. Suppose you are a prosecuting attorney. How has the introduction of PCR changed your job?

2. Why is DNA evidence more useful as exclusionary evidence than for positive identification of a suspect?

3. What difficulties arise in PCR if the DNA that is to be copied is contaminated? What determines if the presence or absence of other DNA sequences in the reaction affects your results?

4. Why are primers usually between 30 and 40 bases long?

5. Why should the G + C content of the forward and reverse primers be similar?

6. What controls the total number of copies of LDH that is amplified with PCR?

Webconnections

For a list of web sites related to the material covered in this chapter, go to **Webconnections** at the *Experiments in Biochemistry* site on the Brooks/Cole Publishing web site. You can access this page at http://www.brookscole.com and follow the links from the chemistry page.

References and Further Reading

Campbell, M. K. *Biochemistry*. Philadelphia: Harcourt Brace, 1998.

Cohen, P. "Ghosts in the Machines." *New Scientist* (1998).

Garrett, R. H., and C. M. Grisham. *Biochemistry*. Philadelphia: Saunders, 1995.

Holland, L. Z., M. McFall-Ngai, and G. N. Somero. "Evolution of Lactate Dehydrogenase-A Homologs of Barracuda Fishes (genus Sphyraena) from Different Thermal Environments." *Biochemistry* 36 (1997).

Jung, H. H., R. L. Lieber, and A. F. Ryan. "Quantification of Myosin Heavy Chain mRNA in Somatic and Branchial Arch Muscles Using Competitive PCR." *American Journal of Physiology* 275, no. 1 (1998).

Oda, R. P., M. A. Strausbauch, A. F. R. Huhmer, N. Borson, S. R. Jurrens, J. Craighead, P. J. Wettstein, B. Eckloff, B. Kline, and J. P. Landers. "Infrared-Mediated Thermocycling for Ultrafast Polymerase Chain Reaction Amplification of DNA." *Analytical Chemistry* 70 (1998).

Promega Corporation. *Technical Manual for pET-5 Expression Vectors*, 1995.

Verhaegen, M., and T. K. Christopoulos. "Quantitative Polymerase Chain Reaction Based on a Dual-Analyte Chemiluminescence Hybridization Assay for Target DNA and Internal Standard." *Analytical Chemistry* 70 (1998).

Voss, K. O., K. P. Roos, R. L. Nonay, and N. J. Dovichi. "Combating PCR Bias in Bisulfite-Based Cytosine Methylation Analysis. Betaine-Modified Cytosine Deamination PCR." *Analytical Chemistry* 70 (1998).